Petit Manan Land Company

Near Bar Harbor, Maine

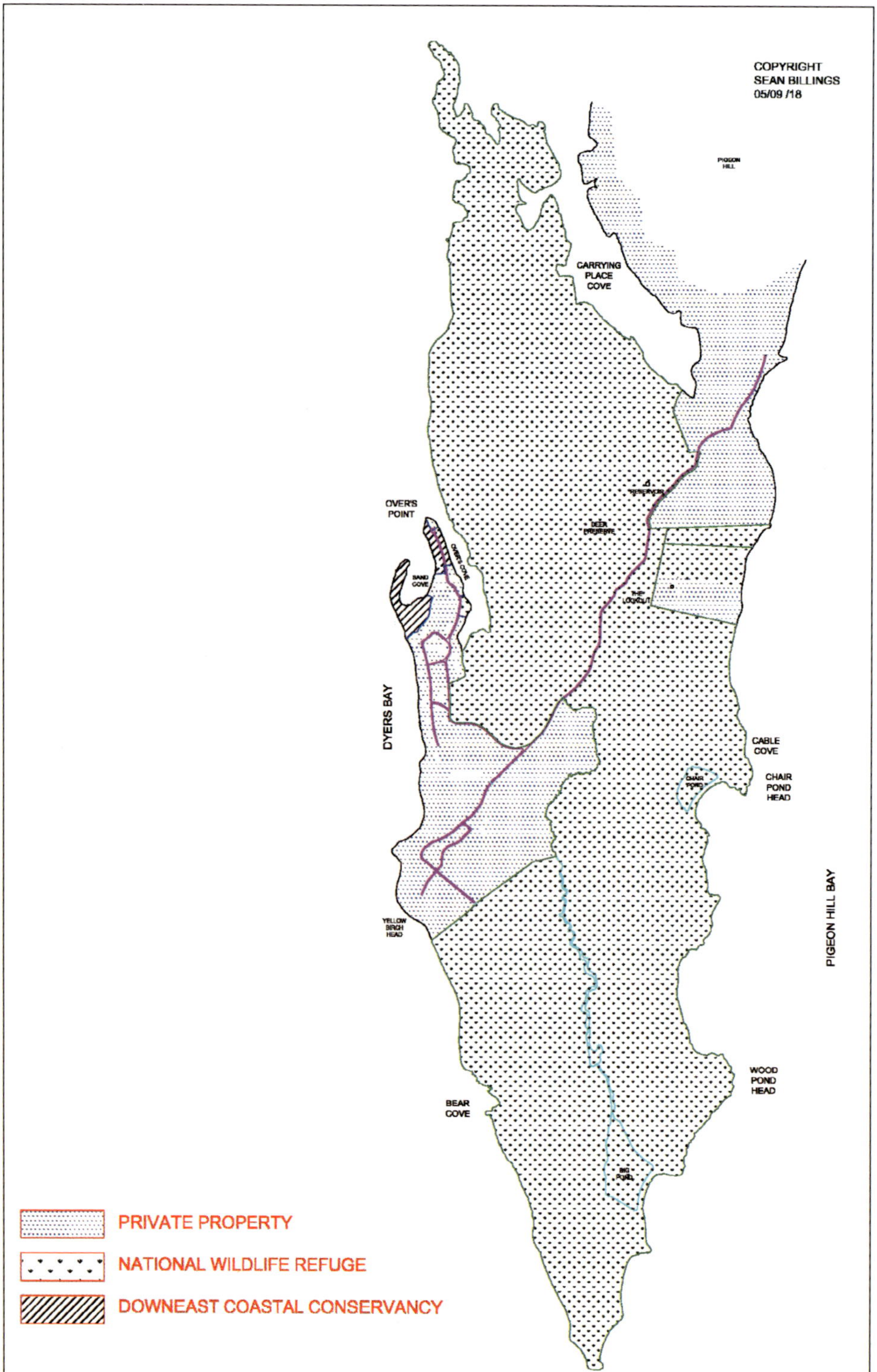
COPYRIGHT
SEAN BILLINGS
05/09 /18
CARRYING
PLACE
COVE
OVER'S
POINT
DYERS BAY
CABLE
COVE
CHAIR
POND
HEAD
PIGEON HILL BAY
YELLOW
BIRCH
HEAD
WOOD
POND
HEAD
BEAR
COVE
PRIVATE PROPERTY
NATIONAL WILDLIFE REFUGE
DOWNEAST COASTAL CONSERVANCY

PETIT MANAN LAND COMPANY

Near Bar Harbor, Maine

THE MAKING OF A NATIONAL WILDLIFE REFUGE

SEAN D. AND JOHANNA S. BILLINGS

AMERICA
THROUGH TIME®
ADDING COLOR TO AMERICAN HISTORY

Frontispiece: Map of Petit Manan Point and the property currently under ownership of the wildlife refuge. The owners of the private property have been generous enough to allow us on their land for research; please show respect and do not trespass.

America Through Time is an imprint of Fonthill Media LLC
www.through-time.com
office@through-time.com

Published by Arcadia Publishing by arrangement with Fonthill Media LLC
For all general information, please contact Arcadia Publishing:
Telephone: 843-853-2070
Fax: 843-853-0044
E-mail: sales@arcadiapublishing.com
For customer service and orders:
Toll-Free 1-888-313-2665

www.arcadiapublishing.com

First published 2020

ISBN 978-1-978-1-63499-212-1

Typeset in 10pt on 13pt Sabon
Printed and bound in England

Preface

The Petit Manan Point section of the Maine Coastal Islands National Wildlife Refuge contains approximately 2,100 acres of land. The refuge currently has two hiking trails—the Birch Point Trail, which goes to Lobster Point and Birch Point on Dyer Bay, and the Hollingsworth Trail, which goes to Chair Pond on the eastern shore overlooking the Petit Manan Lighthouse. Both trails have benches and chairs where visitors can watch birds or the ocean waves. The refuge also contains acres of wild blueberries that can be handpicked when they are in season.

The approximately 175 acres to the east of the refuge is still privately owned. The visitor following the dirt road that goes through the refuge will reach a gate where signage indicates that it becomes private property. Please respect this and do not trespass.

A special thanks goes to Michael and Sandy Denniston for introducing us to the amazing history of Petit Manan Point by taking us on our first tour of the area. Thanks to Barbara and Craig Snapp for the information, photographs, and support. Thanks to Maggi and Jim Kovacs, Frances Mague, Joel Mague, Jessica Mague, Russell Heath, George and Gesele Scully, Claire and Steve Parker, Helen and Jeff Panitt, Mark and Wellner Tremallo, Bonnie Morrison King, Homer Morrison, Bronwen Feeney, Rob Feeney, Joe Shumaker, Anne Dutton, Jack and Lillian Phinney, and Lynn Chase for all the information they provided.

We would also like to thank Julie Ginn and Joy Robinson of the Steuben Town Office for putting up with Sean going through the tax records for days on end, and the staff at the Washington County Courthouse for helping Sean get started when he did not know his way around the deed room.

Thanks also to Sara T. Williams, Linda Welch, Teressa Cultrera, and James Fortier at the Milbridge Office of the Maine Coastal Islands National Wildlife Refuge for access to refuge reports. To anyone else that we may have missed, thank you.

As with any book project, there comes a time when you have to make the decision to publish even though not every detail has been found. There are more areas of Petit Manan to research—the area of Hilton Head, Carrying Place Cove and the Silver Mine needs to be researched—but it was time to publish the book and get what we knew out to everyone.

Early directional sign pointing to Petit Manan in Maine to the left and to the Booth, Denniston, Kent, Tabor, and Whitney cottages to the right.

A current directional sign for the current cottage owners at Petit Manan Point.

This book does not delve into detailed histories of families who built the cottages. Each family could be a book of its own, and we wanted Petit Manan Point to be the star of the book. Chapter 7 does contain background history on the major players involved in the development. We used primary source documents to keep it factual and not just based on memory or hearsay. We hope you enjoy the book as much as we enjoyed learning about this area.

Contents

INTRODUCTION

A visitor today to the Petit Manan National Wildlife Refuge in Steuben Maine would find it hard to believe that the more than 2,100-acre refuge was once destined be the next Bar Harbor. Plans were filed by numerous individuals and companies who tried to sell the nearly 1,300 cottage lots and build hotels and other amenities.

The Petit Manan Land Company started with a grand idea, and by 1896, it had invested money into building roads, a clubhouse, casino, steamboat dock, an icehouse, lumber shed, and a cottage. By 1898, a boarding house and an additional cottage were built. By 1910, a few other private cottages were built, along with a church. But the big money and thriving development eluded company after company for sixty years. With the end of the steamboat era, the dock was the first to disappear. The casino disappeared in the late 1930s. The boarding house, also known as the Smith House or Annex, burned in 1932. The clubhouse finally met its demise in the late 1940s.

In 1975, a large portion of the land was transferred as a partial donation to the Nature Conservancy of the Pine Tree State by Mr. and Mrs. William Mague. It was then transferred to the United States of America in 1976 to become the Petit Manan National Wildlife Refuge. Today, Petit Manan Point is part of the Maine Coastal Island National Wildlife Refuge and remains an important unspoiled stop for migrating birds as well as home to deer, bears, coyotes, porcupines, and other small animals.

1

Early Settlers

The Native Americans of the Wabanaki tribes lived along the Maine coast for thousands of years. No doubt they used the Petit Manan Point in the summer for fishing and hunting, as well as a place to gather berries.

This book documents the modern development; Native American use of Petit Manan Point is beyond its scope. This will be left to another researcher to explore.

From 1794 through 1889, the area was owned by several families and primarily used for sheep farming. Life did not change much until Sylvester K. Abbott got involved with the idea of developing the point into something like Bar Harbor.

Steuben was originally Township 4 in Washington County, Massachusetts. The area was still part of Massachusetts since the state of Maine would not come into existence until 1820. On August 26, 1794, Thomas Ruston, a doctor from Philadelphia, purchased Townships 4, 5, and 6—what are now the towns of Steuben, Harrington, and Addison. On April 22, 1795, Ruston took out a mortgage on Township 4 from Samuel Freeman. On the original map of Township 4, Samuel Freeman Esquire is listed as the agent for the proprietors of the township. At this point, all the land below Carrying Place Cove at Petit Manan Point was intact and not subdivided. On May 25, 1805, Freeman took possession of the property due to breach of the mortgage by Ruston. Samuel Freeman transferred the property to William Freeman on November 2, 1826.

On the Township 4 map that was updated in 1869, six lots are shown as being sold off on Petit Manan Point. At Carrying Place, J. Leighton, Jr., and Moses McCaleb had lots. At Birch Point, H. Leighton, B. Darling, and W. West each had lots, and Reuben Dyer had 100 acres on what is now Over's Point between Pinkham's Cove (now Over's Cove) and Dyer Bay.

Asa Torry, Freeman Torry, and Vinas Torry

On April 27, 1859, Asa, Freeman, and Vinas Torry, all of Steuben, purchased what was known as "The Petit Manan Point lot," containing an estimated 1,673 acres, from

A recent view of the possible foundation of Asa Torrey's house.

Possible foundation of Freeman Torrey's house.

Thomas Palmer of Steuben. The 1861 map from the Washington County Atlas shows A. Torrey having a dwelling at the end of the road near Yellow Birch Head and F. Torrey having a dwelling just up the road from him.

On May 4, 1863, Asa and Vinas Torry took a mortgage on half of the land from Nathan Hinkley of Cherryfield. On November 5, 1863, Nathan and Albion Hinkley of Cherryfield purchased 600 acres of land from Asa and Vinas Torry.

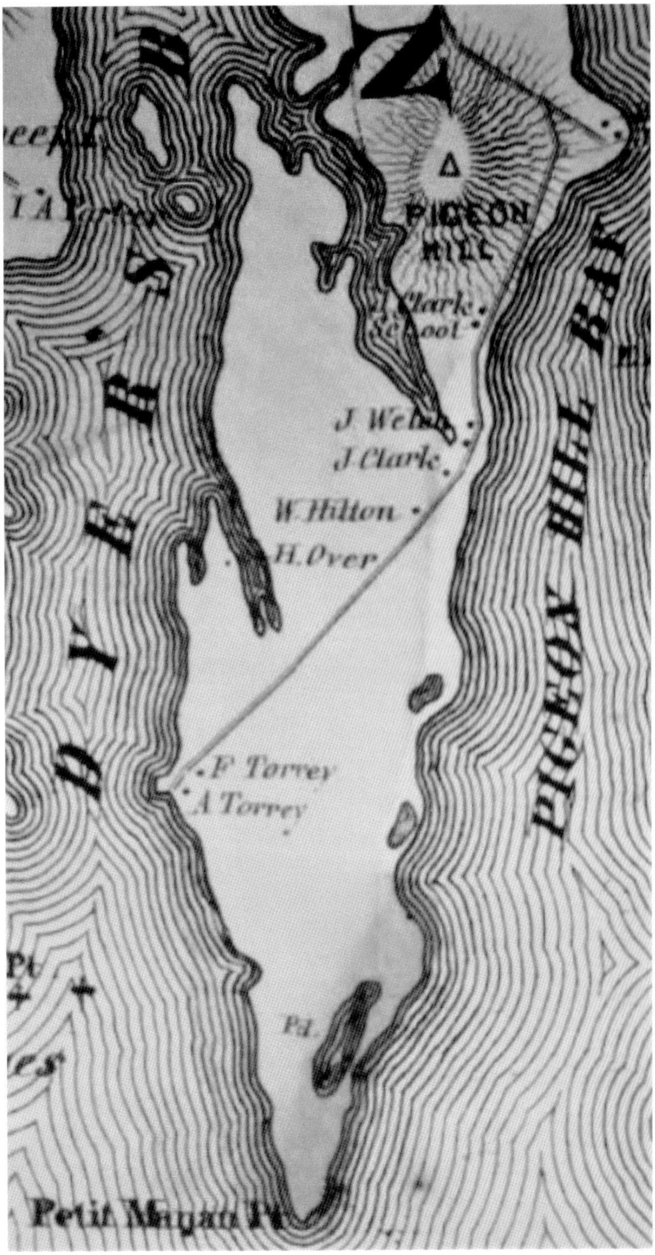

Map of Petit Manan Point from the 1861 Washington County Atlas showing the locations of the Asa and Freeman Torrey homes.

Nathan Hinkley, Albion Hinkley, and Frank Brown

At this point, Nathan and Albion Hinkley owned half of Petit Manan Point and Frank Brown owned the other half. The entire area was used mainly for sheep farming.

Nathan Hinkley is included in the 1850 census of Cherryfield with his wife, Persis, and children, Moses, Leonard, Elias, Joel, Albion, Elizabeth, Handy, Celia, and Angela. Nathan's occupation is listed as farmer. Both Nathan and Albion Hinkley are listed on the 1865 Steuben tax rolls. A note for Albion states "Petit Manan Point, 200 sheep, house and barn." The 1871 through 1873 tax records show Nathan Hinkley paying taxes on 400 acres on Petit Manan Point with half house, quarter barn and shed, and 100 sheep. Albion Hinkley also is shown paying taxes on 400 acres with half house, quarter barn and shed, and 100 sheep. Starting in 1873, both Nathan and Albion are on the tax records for an additional 75 acres purchased from William Freeman. In 1874, the taxed property changes to 400 acres on Petit Manan Point with half house, half barn and shed, and 100 sheep for both Nathan and Albion. In 1875, the property stays the same except both Nathan and Albion increase the sheep herd to 125 each. The 1880 census for Cherryfield shows Albion and his wife, Hannah, living there with their children, Effie, Nathan, Laura, Charles, and Celia. Persis Hinkley is shown on the census as living alone at this time. Albion Hinkley died June 26, 1898. His occupation at the time is listed as trader.

Frank Brown and "others" are listed as the taxpayers for the other half of the point. They paid taxes on 800 acres, one house, half barn and shed, and 200 sheep for 1871 through 1874. In 1875, the taxed property for Frank Brown and others increases to 800 acres, one house, one barn and shed, and 250 sheep.

Birch Point Lot

On August 28, 1857, Miranda P. Stanhope of Old Town sold two Birch Point lots of 50 acres each to Freeman Leighton of Steuben, John McNamara of Steuben, and Warren Leighton of Milbridge. These two lots were the lots previously owned by Handy Leighton and Daniel West on Birch Point. Both Handy Leighton and Daniel West purchased the lots from E. W. Lyman. On February 19, 1863, John McNamara sold his interest in the lots to Warren Leighton. On November 18, 1864, Warren Leighton and Freeman Leighton sold the Birch Point Lot called out as 50 acres to James Mansfield and Nathan Hinkley.

Steuben tax records for 1874 show Daniel West paying taxes on the 100 acres of land belonging to E. W. Lyman, located south of Carrying Place Cove and adjoining the marsh lot on the north. The 1876 tax records show Daniel West paying taxes on the 100-acre E. Lyman lot at Petit Manan Point with a note that it is in dispute with A. Wallace.

James, Greely, Campbell, and Ricker

On September 10, 1874, Nathan and Albion Hinkley sold all the property they owned at Petit Manan Point. Also, on that date, Franklin Brown, James Wallace, and William Godfrey, all of Milbridge, sold all the land they owned below Carrying Place Cove. Both sold to the same group of people: Andrew S. James of Melrose, Mass., Everard Greely of Ellsworth, and David W. Campbell and William C. Ricker, both of Cherryfield. The deed lists the property as being 2,000 acres, including everything from Carrying Place Cove to the point, excluding the Henry Over property of 100 acres. The breakdown of ownership was James one-third, Greely one-third, Campbell one-sixth, and Ricker one-sixth.

Petit Manan Sheep Company and Petit Manan Cranberry Company

By 1876, both the Hinkleys and Frank Brown are no longer on the tax rolls for any property at Petit Manan Point. This same year, the Petit Manan Sheep Company and the Petit Manan Cranberry Company show up for the first time as taxpayers. It is unknown if the Petit Manan Sheep Company and the Petit Manan Cranberry Company were run by James, Greely, Campbell, and Ricker, or if they rented the land to someone else.

In 1876 and 1877, the Petit Manan Sheep Company paid taxes on 1,500 acres of land, two houses, one barn, one sheep shed, 350 sheep, four pigs, and one horse.

The Steuben tax records for 1878 show a Petit Manan Sheep Company paying the tax on 1,500 acres, two houses, one barn, one sheep shed, and an additional 47 acres.

The 1879 tax records show that the 47 acres of land was purchased from F. Sawyer. It shows two houses, one barn, one sheep shed, 425 sheep, and fifteen hogs. The 1880 tax records list the same property but show 500 sheep and fifteen hogs.

The 1881 tax record indicates the Petit Manan Sheep Company paid taxes on 1,600 acres of land, two houses, one barn, one sheep shed, 47 acres of land, and buildings purchased from F. Sawyer. The increase of 100 acres corresponds to the exit of the Cranberry Company from the tax rolls.

From 1876 through 1880, Petit Manan Cranberry Company paid taxes on 100 acres with improvements. The 1881 tax records do not show any property being taxed for the Petit Manan Cranberry Company, and the 100 acres appears to have been added to the sheep company holdings.

Dyer/Over Property

The Over's Point property consisted of 100 acres known as lot 6. It was originally purchased from Township 4 by Samuel Kingsley on June 25, 1795. The land was then sold to Andrew Dyer on June 1, 1799, and then to Reuben Dyer, who bought half of the land on April 12, 1806 and the second half April 1, 1808. Reuben Dyer sold it to Wilson M. Dyer on May 29, 1845.

On August 6, 1850, Wilson M. Dyer sold the lot to Henry Over, an African American who, with his wife, Nancy, had at least two children before they moved to Steuben. Their children were William, Ezra, Albert, Wilbert, Emma, and George. Probably all but William and Ezra were born in Steuben.

George's 1864 birth record shows he was born in Steuben and that his father and mother had both been born in Boston. Henry's occupation was listed as seaman and his color as black. Nancy is listed as "color quadroon," a term which referred to someone being quarter black.

At least two of the Over children fought in the Civil War. William Over enlisted on January 5, 1864. His enlistment papers say he was born in St. George, Maine, and that he was a farmer with a dark complexion. Ezra A. Over enlisted as a private on February 21, 1865, serving until he was honorably discharged September 11, 1865. He was with Company "H," 19th Regiment Maine Volunteers, 1st Heavy Artillery. His birthplace is listed as Rockland, his occupation as seaman, and his complexion as dark.

The 1861 map from the Washington County Atlas shows H. Over having a dwelling near what is now Over's Cove. The 1870 Steuben census records show Henry, Nancy, William, Ezra, and Wilbert Over as living in Steuben. It appears Henry Over died in the mid-1870s, and his property went into the estate.

Steuben tax records from 1871 through 1876 show Henry Over paying taxes on one house, one barn, and a 100-acre home lot. From 1877 through 1881, tax records list the Henry Over Estate as paying taxes on that property. The 1874 tax records also show Ezra Over was paying taxes on a schooner by the name of *Albatross*. On July 7, 1874, the property was transferred to Nancy and Ezra from Henry.

The 1880 Steuben census records show Albert and Wilbert as still living in Steuben, with Albert married to Sarah and Wilbert married to Melissa. Their mother, Nancy, was living in Wilbert's household. Both Albert and Wilbert listed their occupation as fishing. Mrs. Henry Over shows up in the 1882 tax records with one house, one barn, and 32 acres of land. The 1900 Gouldsboro census shows William and Ezra living there with their families.

Foundation on section one, lot 30, that is most likely the Henry Over home site. In 1861, when the National Geodetic Survey was done, the corner of the Overs' barn was used as a reference point. The barn is long gone, but using the coordinates on a current map, they come very close to this foundation. It would then stand to reason that if the Overs' barn was near this foundation, this foundation might be the Overs' house.

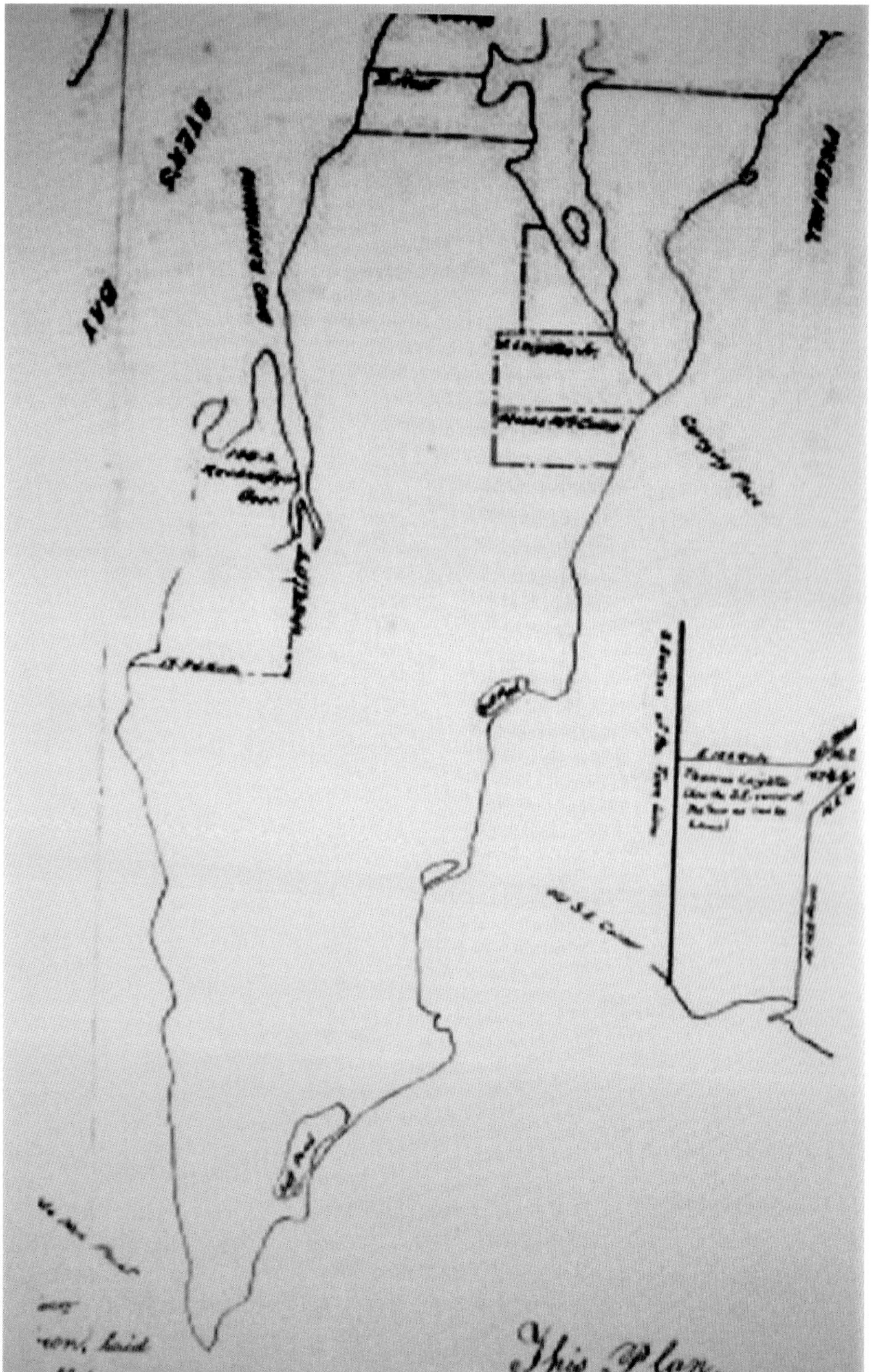

J. D. Parker Map of 1869 showing the location of the Over property.

SYLVESTER K. ABBOTT

On June 3, 1881, William C. Ricker, Emily D. Ricker, Andrew S. James, Ina M. James, Charles Campbell, Frank Campbell, Ella Campbell, Abby W. Campbell, M. W. Patten, F. W. Patten, Everard H. Greely, Ann F. Greely, John W. Hale, Jr., Rebecca W. Hale, George W. Hunt, and Eliza Hunt sold the 2,000 acres they owned at the point to Sylvester K. Abbott.

Was Sylvester K. Abbott planning to develop the point when he purchased all the property? It seems he had some plans beyond sheep farming since he continued to purchase property and purchased the last 100 acres from the Overs to complete the entire tract. He would not have needed these 100 acres if his only intent was to farm sheep.

On August 12, 1881, Ezra Over sold the 100-acre Henry Over property to Abbott. The Overs' 100 acres had been the last piece of land that Abbott needed to bring the entire Petit Manan Point under his control. The deed from Over to Abbott included one exception—it reserved forever a piece of land 40 feet square now surrounded by a fence conveyed and used as a burial lot. This would indicate a burial ground on the point near Over's Point. The location of this burial ground is not known today. Was this the Over or Dyer family plot or something more? None of the land development plans show any burial grounds and none of the journals mention it. It is possible that the Overs exhumed their relatives and reinterned them in Gouldsboro where they eventually ended up. This exception for the burial ground is still in the deed the United States government holds for the Petit Manan National Wildlife Reserve.

With Abbott's purchase of Petit Manan Point, the area changed from sheep farming to commercial development by those hoping to capitalize on the success of nearby Bar Harbor.

A view of the entrance gates looking toward Pigeon Hill; the gates were located just outside the current boundary of the wildlife refuge on Pigeon Hill Road.

2

DEVELOPMENT BEGINS

Before Sylvester K. Abbott purchased the property, Petit Manan Point was used mostly for sheep farms and cranberries. Nothing suggests that anyone was interested in or considered the development potential of the area.

That was about to change. Over the next forty years, several companies would try unsuccessfully to transform the quiet farming area into a bustling community for wealthy tourists.

Sylvester K. Abbott's untimely death was most likely the cause of the failure of the Petit Manan Land Company, drawing different players on to the scene. A number of them were from Belfast, including the Hansons, Duntons, and Kilgores. The Kilgore family would remain involved into the 1930s, with John Kilgore being the last Kilgore brother owning Petit Manan Point at that time.

SYLVESTER K. ABBOTT

It is not clear what Sylvester K. Abbott did first. The initial plans that Frederic R. Page designed were not drawn until July 1889, when the Petit Manan Land Company controlled the land. Did he try his hand at sheep farming initially before deciding to develop the area or was he mining silver?

The 1884 tax records show S. K. Abbott was paying the taxes on half of the 1,600 acres, plus the 100 acres that Over had owned, 47 acres previously owned by Hilton, the Hilton house, half the new house and half the old house, half of two barns, and half the sheep shed. The tax on the other half of the properties was being paid by Charles J. Smith. In 1885, Abbott was paying the taxes on all of the property, and he continued to do so through 1890.

The cover of an early brochure for Petit Manan when it was under development by the Petit Manan Land Company.

Petit Manan Land Company

On April 19, 1889, Abbott transferred the land to the Petit Manan Land Company, a corporation organized in Maine and located in Auburn in Androscoggin County. The deed calls out roughly 2,000 acres, with the 40-foot-square burial lot on the former Over property noted. Also, excepted in the deed are the mine and the right to work the mine that was open there at that time. The June 16, 1889 edition of the *Boston Daily Globe* reports on the developing resort and the silver mine: "10 years ago the gold and silver craze miners sunk a shaft 70 foot deep, widening as they went down. The mine is now filled to within six feet of its top with pure water."

In an early brochure titled *Petit Manan: the new resort on the coast of Maine*, the company address is shown as 11 Otis St., Winthrop Square, Boston. This address is crossed out and "27 School St., Boston," is written in. The directors listed in the brochure are George C. Wing, Auburn, Me., president; C. W. Waldron, New York City, secretary and general manager; S. K. Abbott, Boston, Mass., A, treasurer; T. E. Calvert, Lewiston, Me.; and John L. Fitch, New York City.

It was in July 1889 that the first plans for the point were drawn up by Page, a civil engineer from Malden, Mass. Titled "Plan of the Villa Lots Petit Manan Land Company

Near Bar Harbor Maine," the plans were recorded with Washington County and are still on file there.

The plans consisted of four sections for development along the western shore of the point. The location was probably selected because Green Mountain, or Cadillac Mountain as it is now known, could be seen on the horizon. Most of areas to the north and west were fenced off as a deer park.

Section one was located at the northern tip of the point and the plans showed a proposed steamboat wharf, sand beach, hotel, tennis courts, and 118 villa lots. The casino, clubhouse, steamboat wharf, and a hand full of cottages were eventually built in this section.

Two different plans are on file for section two, located just to the south. The first plan, dated February 1895, contained 148 villa lots. A second plan was for sections two and three and dated April 1895. It shows an additional 376 villa lots in the second section, for a total of 524 in section two.

Section three also has two different sets of plans on file. The first plan, dated July 1889 and revised in February 1895, shows 338 villa lots. The second plan for section two and three, dated April 1895, shows an additional eighty-three lots, along with a site for a hotel.

Section four, located at the far southern tip of the point, was laid out to contain 305 villa lots and an ocean-view drive along the coast.

The Deer Park

From the beginning, the deer park was part of the general allure and intrigue, reflected by how newspapers reported on it.

In the June 16, 1889 edition of *Boston Daily Globe*, the main headline was "Rival of Bar Harbor" with a subhead of "Deer Park of Five Hundred Acres to be Stocked."

The December 9, 1889 *Bangor Daily Whig and Courier* ran a story with the headline "Immense Deer Park." A portion of the article reads:

> One of the features of the new summer resort of the Petit Manan Land Company, near Bar Harbor, will be a large deer park well stocked with deer. It will be the largest deer park in the United States. Wire netting to enclose the park has just gone over the Maine Central from Gilbert & Bennett, Georgetown, Conn., on the way to Hancock Point. This netting is four feet wide and eight miles long and weighs many tons. It will be put up double so it will be eight feet high and four miles around the park, which will include 700 acres.

The deer park was of great interest at the time, and story after story appeared in the papers. An article in the January 29, 1890 edition of the *Bangor Daily Whig and Courier* states:

> A gentleman who returned from Washington County describes the deer park which is being established on Petit Manan Point by a Boston company, of which Mr. Abbot of Boston is president. A section of the point of about 800 acres has been fenced off by a fence eight feet high and from four to five miles in length. It is to be stocked with deer for private sport, the first animal to be introduced into the enclosure in about two weeks.

About two weeks later, the February 13, 1890 edition of the *Bangor Daily Whig and Courier* reported:

> The Boston people who are arranging to inaugurate a deer park on Petit Manan Pond [sic.] are making good progress in their enterprise. A section of the point comprising about 800 acres has been fenced off by a fence eight feet in height, and from four to five miles in length. It is to be stocked with deer for private sport, the first animal to be introduced into the enclosure in about a week.

Three months later, the May 13, 1890 edition of *Bangor Daily Whig and Courier* reported:

> One of the most expensive enterprises ever undertaken in Maine is the deer park at Petit Manan Point. Thirty deer have been obtained for this fine resort, which is destined to be one of the most popular on the Maine coast. The park embraces about 300 acres, is enclosed with a high wire fence, and is controlled by a rich company who are right in believing that they possess one of the finest tracts of land on the entire Atlantic coast.

It appears that S. K. Abbott died in 1890. The 1893 and 1894 tax records show his estate paying the taxes on the 1,600 acres of land, 100-acre former Over place, 47-acre former Hilton place, two barns, one house, one hall, and one sheep shed. It is unknown if his death played a role in the demise of the Petit Manan Land Company, but in 1895, the Petit Manan Land and Industrial Company became the owner of the property.

A view of the Deer Park fence.

Whig and Courier.

MONDAY, DECEMBER 9, 1889.

[For additional Despatches see First Page.]

BY TELEGRAPH.

MAINE NEWS.

IMMENSE DEER PARK

To Be One of the Features of a Maine Resort.

PORTLAND, Dec. 7. One of the features of the new summer resort of the Petit Manan Land Company, near Bar Harbor, will be a large deer park well stocked with deer. It will be the largest deer park in the United States. Wire netting to inclose the park has just gone over the Maine Central from Gilbert & Bennett, Georgetown, Conn., on the way to Hancock Point. This netting is four feet wide and eight miles long and weighs many tons. It will be put up double, so that it will be eight feet high and four miles around the park, which will include over 700 acres.

Right: An article about the Deer Park at Petit Manan from 1889 showing how much interest there was in the Deer Park.

Below: A recent photo of wire left from the Deer Park fence.

PETIT MANAN LAND AND INDUSTRIAL COMPANY

On November 21, 1895, the property was transferred to the Petit Manan Land and Industrial Company from the Petit Manan Land Company. In the deed, the Petit Manan Land Company is listed as having its principal place of business in Belfast, Waldo County. To purchase the land, the Petit Manan Land and Industrial Company gave a promissory note of $100,000 payable before ten years, plus capital stock in the Petit Manan Land and Industrial Company in the amount of $402,275. The Petit Manan Land and Industrial Company assumed all liabilities of the Petit Manan Land Company. Its president was Edgar F. Hanson and the treasurer was Gustavus C. Kilgore.

On December 26, 1895, the Petit Manan Land and Industrial Company entered an indenture with the International Trust Company for 700 acres.

The October 8, 1895 edition of the *Bangor Daily Whig and Courier* reported that the Petit Manan Land and Industrial Company had been organized in Belfast for the purpose of buying, selling, and renting lands with $2,500,000 capital stock, of which $1,080 was paid in. The officers were President A. F. Hanson and Treasurer G. C. Kilgore, both of Belfast. The certificate was approved October 5, 1895.

The November 11, 1895 edition of the *Bangor Daily Whig and Courier* stated that James T. Pottle of Belfast "has drawn the plans and is figuring on the erection, at Petit Manan Point, of a $25,000 hotel. The building, if erected, will be owned by the company and other parties. If built, Mr. Pottle will take down a crew this fall."

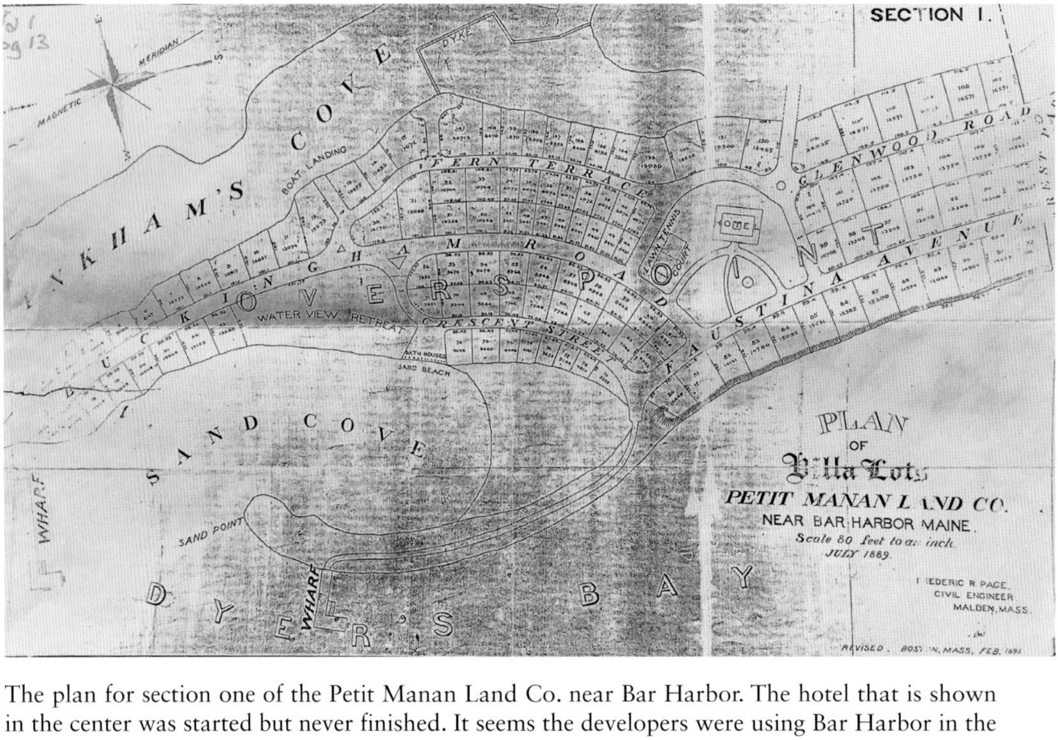

The plan for section one of the Petit Manan Land Co. near Bar Harbor. The hotel that is shown in the center was started but never finished. It seems the developers were using Bar Harbor in the title as a possible selling feature.

Artist's drawing of the proposed hotel from the Petit Manan Land Company brochure.

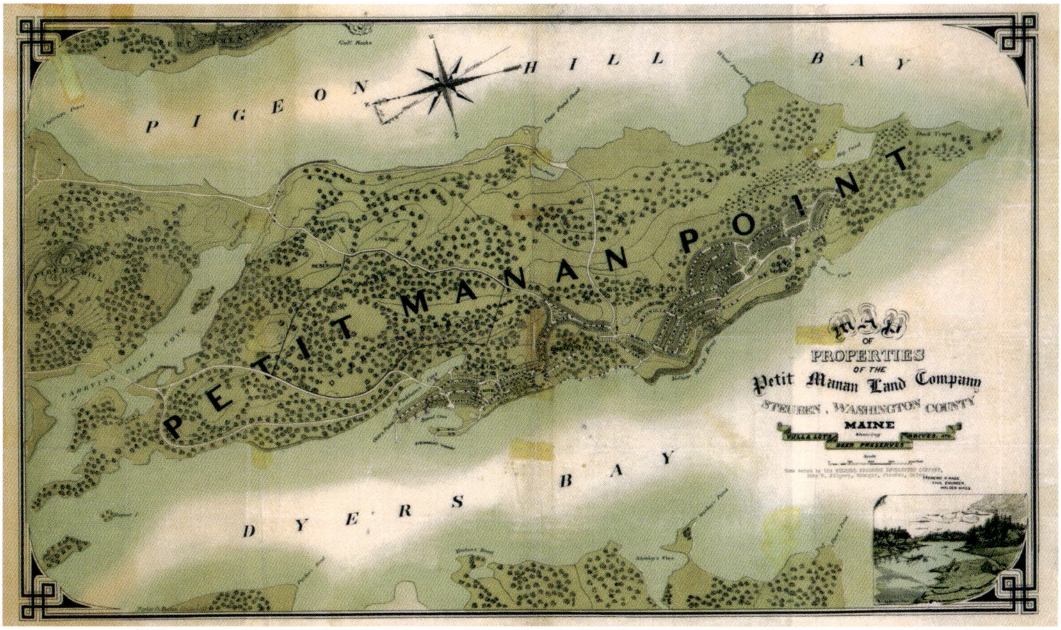

One of the original promotional plans for the Petit Manan Land Company.

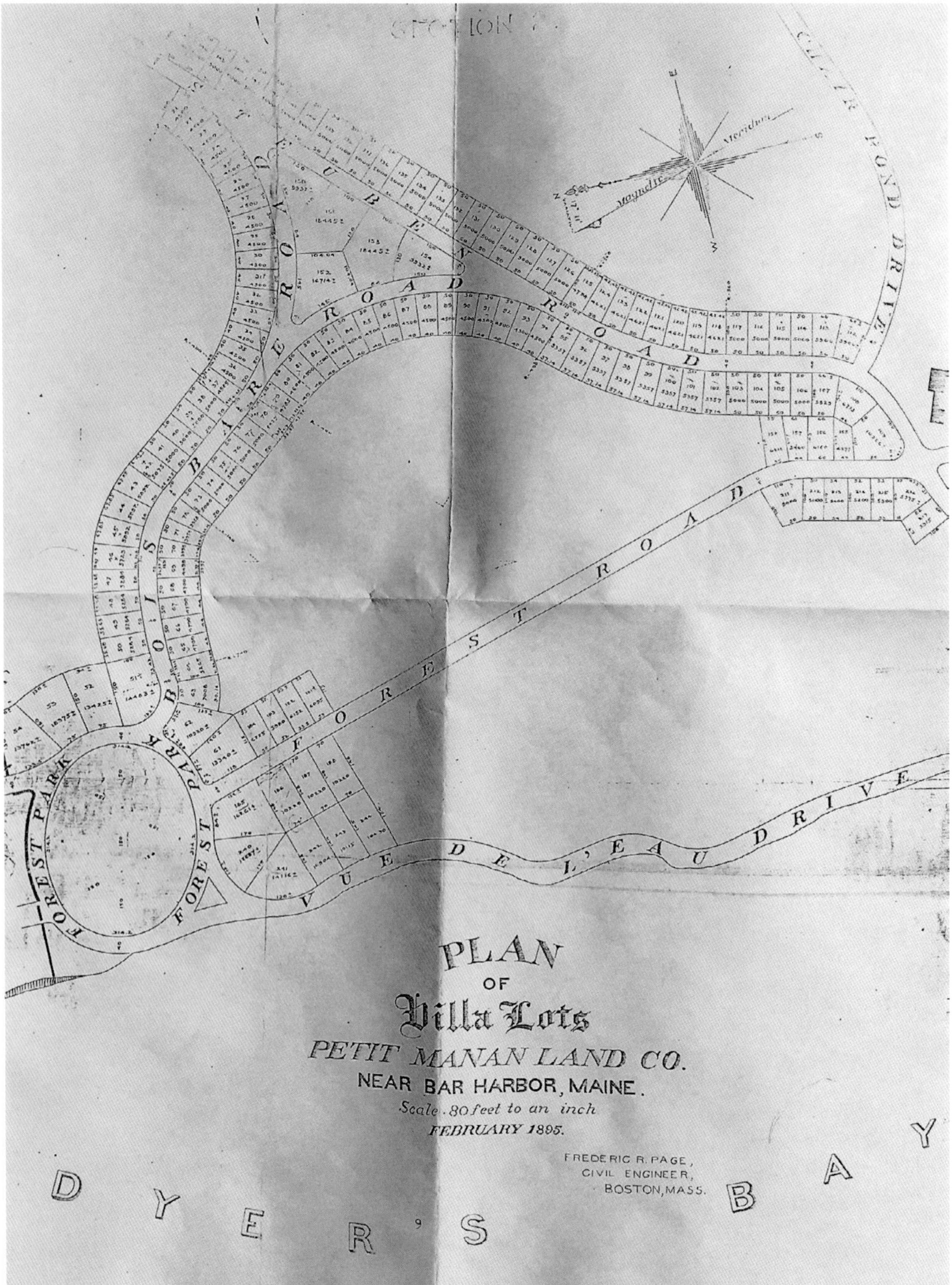

The plan for section two.

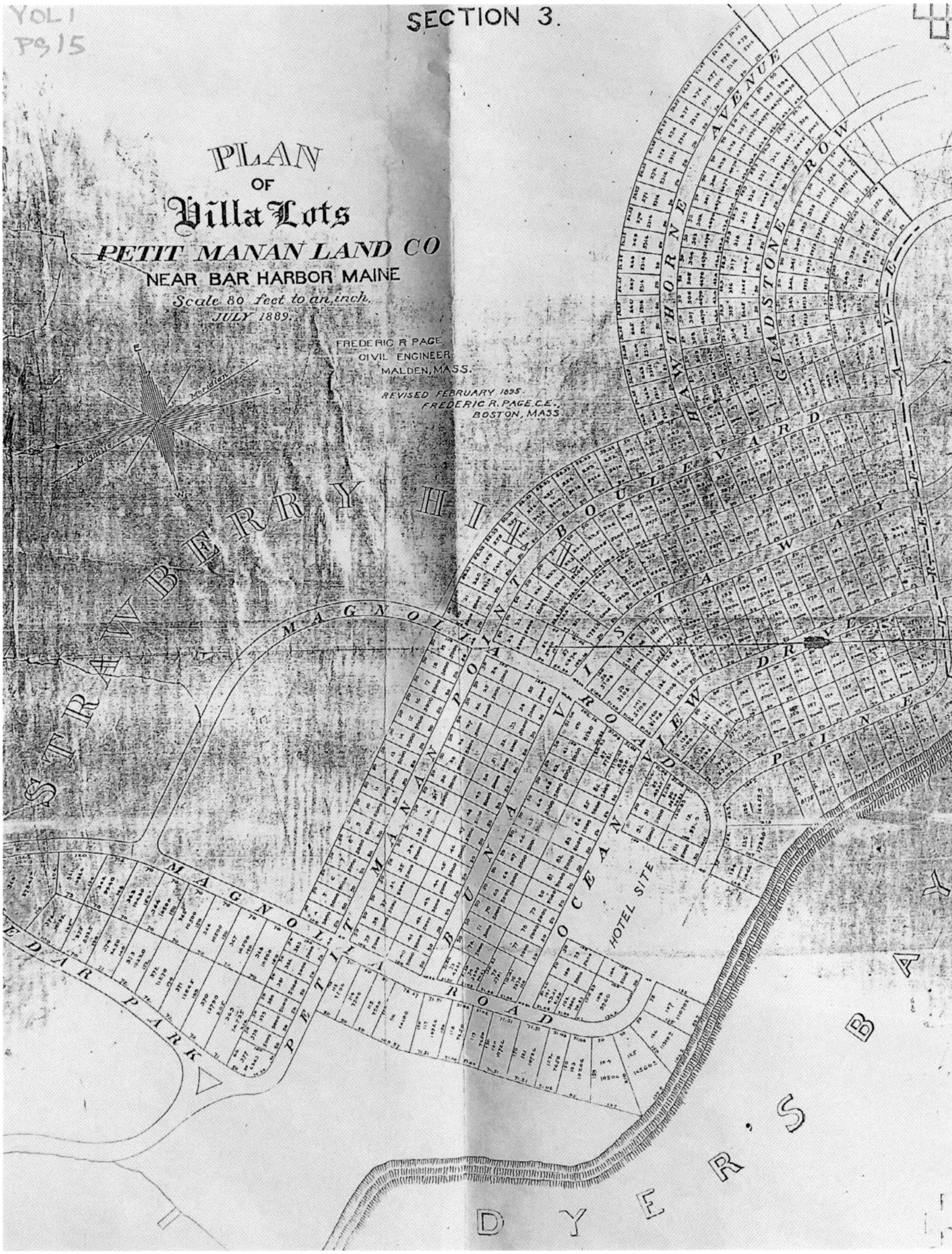

The plan for section three.

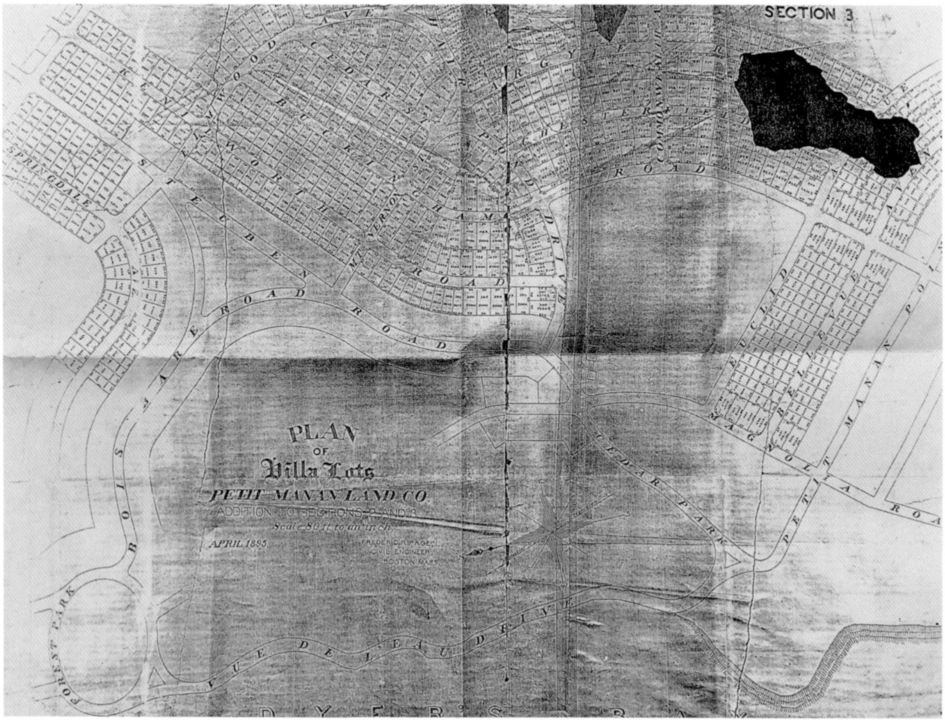

Above: The revised plan for sections two and three.

Below: The plan for section four.

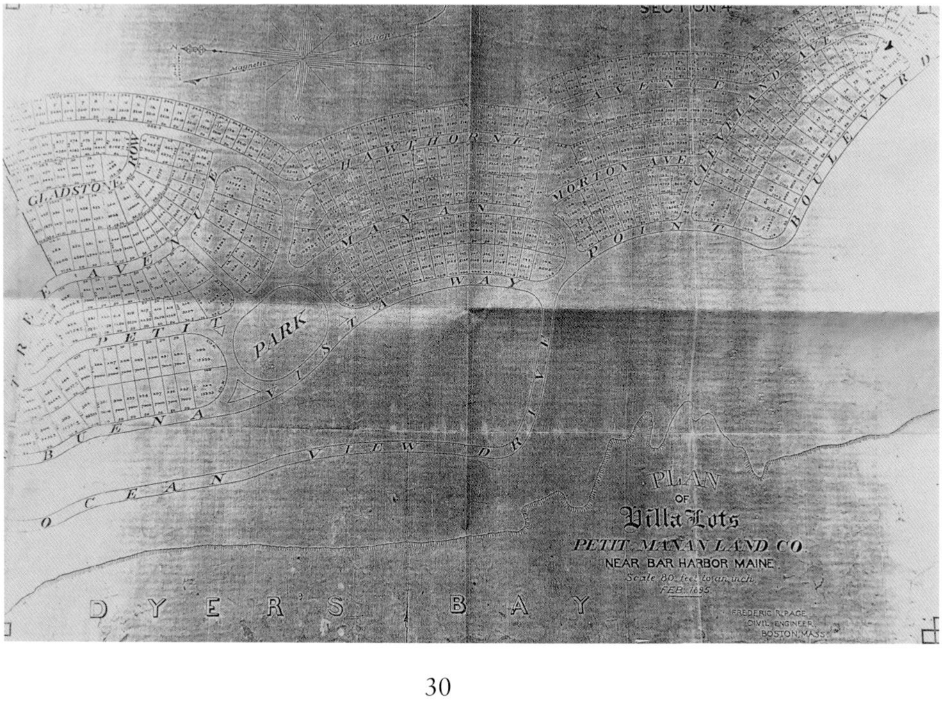

While the Petit Manan Land and Industrial Company was moving forward with development, it was still dealing with past issues inherited from the Petit Manan Land Company. An article from the *Bangor Daily Whig and Courier*, dated October 27, 1896, illustrates this:

> Hearing at Augusta on application for the Receiver for Petit Manan Land Company Augusta, Me., Oct. 26 A hearing was held before Judge Whitehouse today on a petition for the appointment of a receiver for the Petit Manan Land Company. The judge is holding the question under advisement.

The same newspaper reported on February 19, 1897:

> A hearing was held Wednesday at the courthouse at Augusta of the creditors of the Petit Manan Land Company before Joseph Williamson Jr. as master in chancery, and bills amounting to about $20,000 against the company were received and allowed. Hon. S.S. Brown was counsel for the receiver. There were present Hon. E.F. Hanson of Belfast. Hon. R.W. Dunn of Waterville, Dr. G.C. Kilgore of Belfast, Dr. W.G. Fuller of Pittsfield, J.H. Flannigan of Portland and Capt. A. Parker of Steuben.

Another article, dated June 8, 1897, reads:

> Augusta, Me., July 7. A hearing was held in Augusta today in the case of the Petit Manan Land Company vs. the Petit Manan Land and Industrial Company before Joseph Williamson Jr. as master in chancery, and claims amounting to $5,000 were presented and allowed, making $48,000 up to date.

Another portion of the same article reads:

> Another hearing of Petit Manan Land Company vs. Petit Manan Land and Industrial Company was held before Judge W.P. Whitehouse, a case in equity on a supplemented bill filled by Eben W. Freeman, of Portland, attorney for Annie W. Davis, of Boston, a stockholder in the Petit Manan Land and Industrial Company The bill prayed that the decree appointing a receiver might be admitted to answer. The decision was reserved.

Within a very short period of time, the Petit Manan Land and Industrial Company found itself not doing much better than the Petit Manan Land Company had. By 1898, the property again changed hands, going to Henry Kilgore and Frank Wilson.

Henry Kilgore and Frank Wilson

At a sheriff's sale on November 9, 1897, the sheriff of Washington County, Isaac Longfellow, sold the Petit Manan Land and Industrial Company to John H. Flannagan for $2,528.54. Edgar F. Hanson was president at the time and R. W. Dunn was appointed receiver by the court. On March 23, 1898, Dunn sold the property by court order to Henry Kilgore, brother of Gustavus Kilgore, and Frank Wilson for $1,551.95. The

Kilgore family would play a prominent role in Petit Manan Point, with another brother, John M. Kilgore, and sister in-law, Abbie O. Kilgore, eventually being involved as well.

The deed to Henry L. Kilgore says that it includes all the deer contained in the deer park and all the plans, pamphlets, stationary, case of shelves, and other property stored in the stable of G. C. Kilgore in Belfast.

The March 14, 1898 edition of the *Boston Daily Advertiser* reported that, on March 12, "the property of Petit Manan Land and Industrial Company was sold at auction this afternoon for $11,500 beyond liabilities. The proceeds will be divided among the creditors."

In a strange twist of fate, Henry Kilgore died on December 4, 1898, and the project was once again in legal limbo.

THE MAINE COAST CLUB

On December 19, 1900, the property was transferred from the Estate of Henry L. Kilgore, Frank I. Wilson, Charles A. Wilber, and Everard H. Greely and consolidated under Robert F. Dunton, trustee. The Maine Coast Club was set up with seven directors; each director was to get a lot—as was G. C. Kilgore. For every share of common stock purchased by a member, the purchaser got one lot of land. By this point, the property still had debt, as follows:

First National Bank of Ellsworth	$4,433.94
E. H. Greely	$252.20
First National Bank of Madison	$7,445.79

A share of the Maine Coast Club cost $500, and from each sale, $150 was to go to the First National Bank of Ellsworth and E. H. Greely, to split, and $150 was to go to First National Bank of Madison. From each sale, 10 percent was to be held by the trustees for incidental expenses under the direction of Abbie O. Kilgore, Georgie G. Hanson, and J. M. Kilgore.

In addition, the People's National Bank of Belfast held three notes for a total of $2,675. The People's National Bank of Searsport held a note for $750, and a $1,391.30 salary was owed to J. J. Grant for caring for the property since the receiver's sale.

After the debts were paid, 80 percent of the sales were to go to Abbie O. Kilgore and Georgie G. Hanson until they each received $15,000. The other 20 percent was to go to J. M. Kilgore. After that, any shares not sold were to become the property of Abbie O. Kilgore, Georgie G. Hanson, and J. M. Kilgore.

The New York Times reported on the new Maine Coast Club on March 17, 1902:

> The Maine Coast Club. Millionaires, Artists and Brokers to Build a "Sportsman's Paradise." The Maine Coast Club is a new organization of millionaires, artists and brokers, which proposes to establish a sportsman's paradise on Petit Manan Island [sic.], Maine, 12 miles from Bar Harbor. Each member of the club, upon being voted in, receives a deed to a lot of land at the time of paying his initiation fee, and 100 cottages are to be erected the coming season. The club is managed by an executive directorate,

which includes J.W. Paul of the Philadelphia Art Club, W.F. Owens, New York; Herbert Jacques, Boston; C. Vey Holman, Bangor; E.H. Dennison [sic.], Philadelphia; J. Frank Turner, Plymouth, Penn., and Charles B. Foltz, Philadelphia.

The 100 cottages mentioned in the article were never built and within six years, by December 31, 1906, the Maine Coast Club was behind on payments to the First National Bank of Madison. The bank sent a request to transfer the property to Robert F. Dunton for breaking the trust made on December 19, 1900. That trust stated that the First National Bank of Ellsworth, the First National Bank of Madison, and Everard H. Greely were to be paid within two years of December 19, 1900.

By this time, John M. Kilgore had taken over the debt from Everard H. Greely and the First National Bank of Ellsworth, so all the debt was held by him and the First National Bank of Madison.

On January 21, 1907, Robert F. Dunton, trustee, transferred the property to Henry H. Gray and Charles H. Clark in trust for the First National Bank of Madison and John M. Kilgore. At this time, the First National Bank of Madison went into liquidation and its assets were transferred to the Augusta Trust Company.

In July 1907, John M. Kilgore purchased the debt from the First National Bank of Madison and the property was transferred to Frank W. Sawyer, trustee, for John M. Kilgore. On November 19, 1908, John M. Kilgore became the sole owner after Abbie O. Kilgore, Georgia G. Hanson, Gustavus C. Kilgore, and Edgar F. Hanson deeded their portion over. On December 10, 1908, Frank W. Sawyer transferred everything over to him.

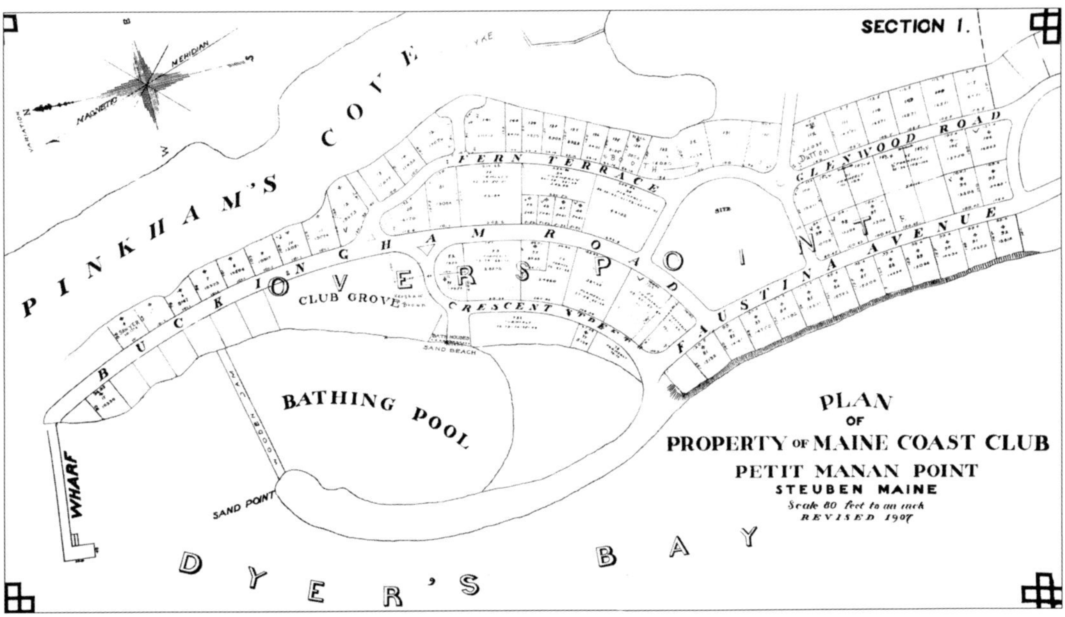

The revised section one plan from the Maine Coast Club.

John M. Kilgore

From 1908 to the 1920s, not a lot of activity appears in published reports or records, and Kilgore ran the property as a club. He continued to call it the Maine Coast Club but also used the names Kilgore and Kilgore Seashore Development Company. A brochure titled *Kilgore* lists hunting, fishing, bathing, horseback riding, tennis, dancing, and bowling as the activities available. The clubhouse was turned into a lodge for guests and a log cabin was also available. The rate was advertised as $4 per day.

On January 4, 1922, Kilgore indentured all the blueberries on the property for ten years to be delivered to A. L. Stewart and Sons in Cherryfield. On January 10, 1922, Kilgore took out a $15,000 mortgage on the land from Charles A. and Frank H. Stewart of Cherryfield. On May 25, 1923, Kilgore took out a second mortgage on the property for $10,000 from Harry W. Goddard. Then, again, on February 8, 1924, he took out an additional mortgage for $5,000 from Goddard. These were in addition to the mortgage he had taken out from Charles A. and Frank H. Stewart on January. 10, 1922. The Goddard family often spent time at Petit Manan Point, using the clubhouse after John M. Kilgore turned it into a lodge.

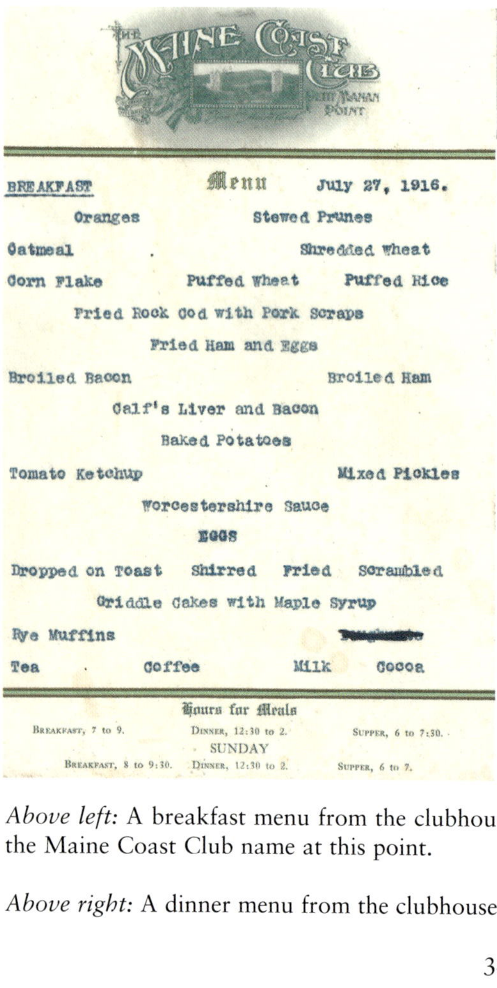
The Maine Coast Club
Petit Manan Point

BREAKFAST — Menu — July 27, 1916.

Oranges — Stewed Prunes
Oatmeal — Shredded Wheat
Corn Flake — Puffed Wheat — Puffed Rice
Fried Rock Cod with Pork Scraps
Fried Ham and Eggs
Broiled Bacon — Broiled Ham
Calf's Liver and Bacon
Baked Potatoes
Tomato Ketchup — Mixed Pickles
Worcestershire Sauce
EGGS
Dropped on Toast — Shirred — Fried — Scrambled
Griddle Cakes with Maple Syrup
Rye Muffins
Tea — Coffee — Milk — Cocoa

Hours for Meals
Breakfast, 7 to 9. Dinner, 12:30 to 2. Supper, 6 to 7:30.
SUNDAY
Breakfast, 8 to 9:30. Dinner, 12:30 to 2. Supper, 6 to 7.

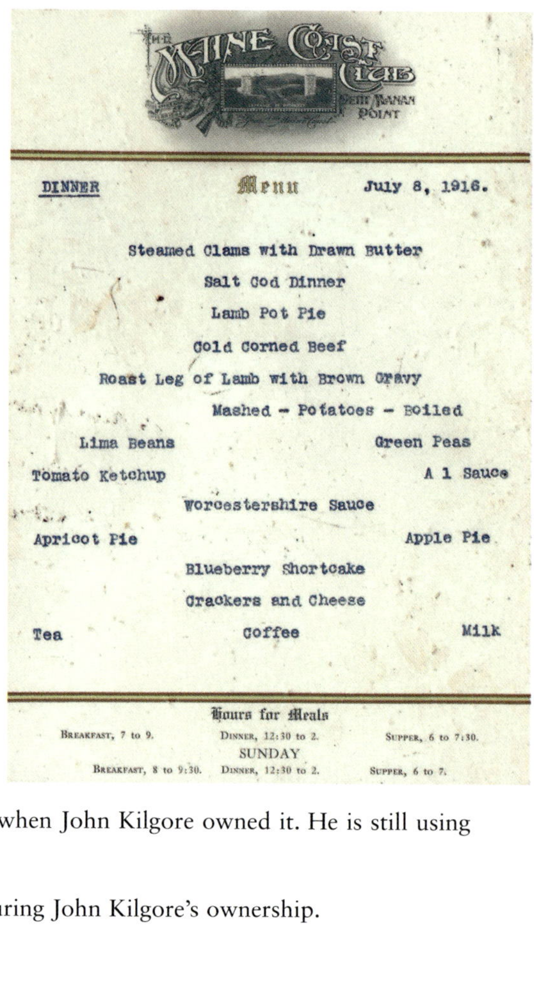
The Maine Coast Club
Petit Manan Point

DINNER — Menu — July 8, 1916.

Steamed Clams with Drawn Butter
Salt Cod Dinner
Lamb Pot Pie
Cold Corned Beef
Roast Leg of Lamb with Brown Gravy
Mashed - Potatoes - Boiled
Lima Beans — Green Peas
Tomato Ketchup — A 1 Sauce
Worcestershire Sauce
Apricot Pie — Apple Pie
Blueberry Shortcake
Crackers and Cheese
Tea — Coffee — Milk

Hours for Meals
Breakfast, 7 to 9. Dinner, 12:30 to 2. Supper, 6 to 7:30.
SUNDAY
Breakfast, 8 to 9:30. Dinner, 12:30 to 2. Supper, 6 to 7.

Above left: A breakfast menu from the clubhouse when John Kilgore owned it. He is still using the Maine Coast Club name at this point.

Above right: A dinner menu from the clubhouse during John Kilgore's ownership.

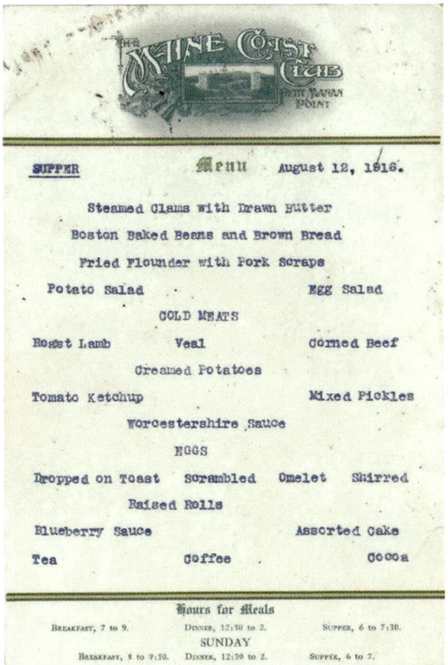
The Maine Coast Club
Petit Manan Point

SUPPER Menu August 12, 1916.

Steamed Clams with Drawn Butter
Boston Baked Beans and Brown Bread
Fried Flounder with Pork Scraps
Potato Salad Egg Salad
COLD MEATS
Roast Lamb Veal Corned Beef
Creamed Potatoes
Tomato Ketchup Mixed Pickles
Worcestershire Sauce
EGGS
Dropped on Toast Scrambled Omelet Shirred
Raised Rolls
Blueberry Sauce Assorted Cake
Tea Coffee Cocoa

Hours for Meals
Breakfast, 7 to 9. Dinner, 12:30 to 2. Supper, 6 to 7:30.
SUNDAY
Breakfast, 8 to 9:30. Dinner, 12:30 to 2. Supper, 6 to 7.

Above left: A supper menu from the clubhouse during John Kilgore's ownership.

Above right: The front cover of a brochure for the "Kilgore," one of the other names John Kilgore used for the resort.

Penzance in Maine

On May 29, 1926, John M. Kilgore sold the property to Ferdinand H. Shumaker and held the mortgage himself for $27,500. It appears that Shumaker tried to turn the property into more of a country club with the addition of golf and a new name—Penzance-in-Maine Country Club. Notes made by Edward Evans "Ted" Denniston in 1994 state that Shumaker started a golf course on the way out to Bear Cove but completed only one tee. Shumaker did not make it very long, and on February 16, 1929, John M. Kilgore foreclosed and took possession of the land again. Even though John M. Kilgore sold the property to Shumaker, he still had the mortgages against the property from the Stewarts. On February 23, 1928, the Stewarts filed a notice of foreclosure against Kilgore. The foreclosure reserved the right of John M. Kilgore to sell and convey the platted lots by the rules of the Kilgore Seashore Development Company. In 1926, John M. Kilgore also bought back a number of lots from people, as well as from the inhabitants of Steuben lots that had been lost for back taxes. Before the foreclosure went through, the Stewarts sold the mortgage on January 10, 1929 to the Mechanics National Bank of Worcester. On January 19, 1929, John M. Kilgore took out a mortgage from Minnie C. Casey against a number of the individual lots. John M. Kilgore died in 1931, and in his will, he gave any lots he owned at Petit Manan Point to Minnie C. Casey. It is unclear if Casey ever took position of any of the lots or if she was involved at all after this point.

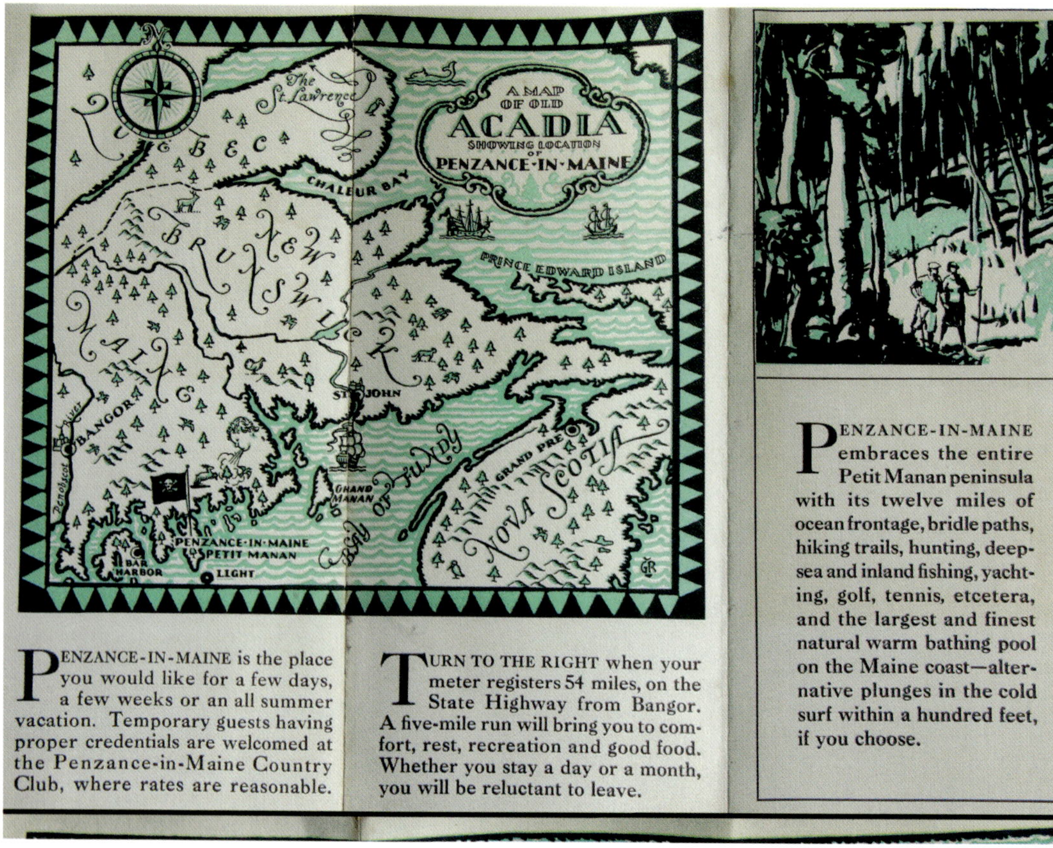

The front of a brochure for Penzance-In-Maine, the name used during Ferdinand Shumaker's brief ownership.

PETIT MANAN IN MAINE INC.

With John Kilgore's death, the foreclosure by Mechanics National Bank of Worcester, as well as mortgages held by Harry Goddard, meant that the property was in a tangle of debt. Harry Goddard also died prior to 1930, so his daughter, Eleanor G. Daniels, became involved.

Eleanor G. Daniels set up Petit Manan in Maine Inc. to serve as a family vacation retreat, with Winthrop G. Hall managing it. In 1941, Winthrop G. Hall was listed as treasurer, Madeline T. Hall as acting clerk and director, and Eleanor G. Daniels as director and stockholder.

On January 25, 1930, Mechanics National Bank of Worcester sold lots 76 and 77 in section one and lots 145 to 150 in section three to Petit Manan in Maine. Then, on December 15, 1930, it sold lots 113, 114, and 116–120 in section three to the same entity. Petit Manan in Maine Inc. sold lots 76 and 77 in section one to Emma M. Sibley, lots 145 to 150 in section three to Leon F. Lawrence, and lots 113 to 120 in section three to Robert Simpson. The bank then held the bulk acreage of the land for ten years.

The Mechanics National Bank of Worcester and Harry Goddard mortgages both excluded the plotted lots that John M. Kilgore reserved the right to sell.

Petit Manan in Maine Inc. was centered around the farmhouse with ten cabins and a community house. The cabins were modular and made of 6 × 10-foot panels ordered from Greensville. To add additional bedrooms, a double cabin was built off the original designs. The cabins were powered by a direct current (DC) generator. Commercial electrical power was not run down to Petit Manan Point until 1948.

During the time the Halls were operating Petit Manan in Maine, some nearby residents were employed by them, including Jack Phinney who lived near Carrying Place Cove. He worked for the Halls, weeding gardens, cutting wood, and painting the cabins. Phinney's grandmother, Lucy, did laundry for people down at the point.

In 1930, Petit Manan in Maine was paying the taxes on the icehouse, farmhouse, barn, sheep shed, stable, hog house, shop, log cabin, two henhouses, three pumping stations, garage, ten cabins, and the community center. This left John M. Kilgore paying the taxes on the casino, clubhouse, Smith House (Annex), Smith stable, and the Gougins stable. John M. Kilgore died in 1931 and his estate held the properties until around 1938.

On August 26, 1939, Petit Manan in Maine Inc. purchased twenty-five lots in section one, fifteen lots in section two, nine lots in section three, and one lot in section four from the Town of Steuben. These were the lots that had been taken by the town over the years for taxes.

On April 16, 1940, Eleanor G. Daniels purchased the majority remaining property from Mechanics National Bank of Worcester.

On January 24, 1941, the lots that had been purchased from the town were transferred to Daniels for consideration of debts she was owed by Petit Manan in Maine Inc.

At this point, Daniels owned most of the original property. Between 1941 and 1950, she sold a few lots. On January 31, 1950, she transferred 360 acres, including the clubhouse and two log cabins, to the Katahdin Council Inc. of the Boy Scouts of America. On February 28, 1950, she transferred 500 acres near Bear Cove to the Maine Department of Inland Fisheries and Game. On March 7, 1950, she transferred most of the rest of the land to Petit Manan in Maine Inc.

On March 6, 1950, Daniels sold the farmhouse and 150 acres around it to William R. Mague, husband of one of the Hall daughters, Priscilla.

The sale of the farm excluded the buildings that belonged to Petit Manan in Maine Inc. Those buildings would later be sold off and used as cottages. Eight of the original ten cottages are still in use, having been moved to various locations. Cabins 1 and 3 were sold to the Heaths; Cabins 2 and 8 and the community building were used to make a cottage known as Stornoway; Cabins 6 and 7 were used to make the Reussers cottage; and cabins 9 and 10 were sold to the Skinners and moved to Yellow Birch Head before being sold to the Humphreys and moved to Carrying Place Cove. Richard and Muriel Humphrey had honeymooned at Petit Manan in Maine in 1947 and, in the 1970s, they purchased the cabins and moved them. The Humpreys' children still own them. Cabins 4 and 5, which were given to the Boy Scouts, are the two that are lost.

The deed to the Boy Scouts includes a clause stating that with a thirty-day notice, the land would be abandoned and then deeded back to Daniels within sixty days. The deed to the Maine Department of Inland Fisheries and Game did not contain an abandonment clause like the Boy Scout deed, but both properties were transferred back to Eleanor G. Daniels. The Boy Scout property was transferred on May 27, 1958 and the Maine Department of Inland Fisheries and Game property on March 26, 1958.

On February 17, 1959, the land that was transferred back by the Boy Scouts and the State of Maine was transferred to Petit Manan Industries, a company organized by William R. Mague.

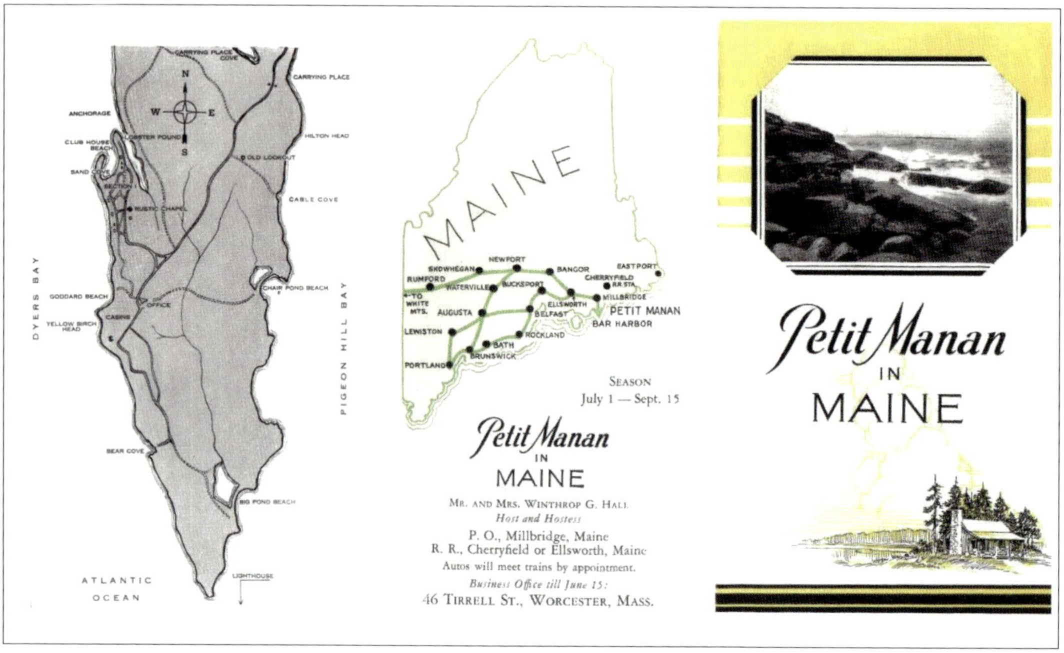

A brochure for Petit Maine in Maine during time the Halls operated it.

Opposite above: A postcard view of a cabin at Petit Manan in Maine, most likely cabin 1.

Opposite below: A postcard view of cabin 9 and cabin 10 at Petit Manan in Maine, the hoop in the lower center is possibly for a mini-golf course setup in the fields in the 1930s.

Residence is enjoyed in individual Log Cabins, located here and there along the edge of the woods; near enough for neighborliness but separated enough for privacy. Each affords beautiful views of ocean, woods and mountains.

A nice place for families with children or for ladies leisuring alone.

Honeymoon parties come to Petit Manan each year because friends have recommended it as ideal for that purpose.

Single Cabin

Indoor affairs are made possible and pleasurable at the well-appointed Log Lodge with its open fire, mellow lights, inviting furniture, piano and games.

Library End of Lodge

Community Church
All-Souls-By-The-Sea

Petit Manan
IN
MAINE

OCEAN - WOODS - LAKES

2500 acres of natural Maine woodland and open spaces on a peninsular, just off U. S. Highway No. 1.

There is health in the invigorating, ozone-laden air. Nowhere is slumber more refreshing.

Twelve miles of varied shore front to invite exploration and picnics. Rugged rocks and spectacular surf on the side of the open ocean. Sandy beaches and quiet water, for bathing, on the Bay side.

Tennis, lawn golf, badminton, boating, fishing, camping, mountain climbing, informal parties of great variety.

Golf links within easy driving distance. Many points of historic interest.

RELIEF FROM HAY FEVER.

Corner of Glassed-in Dining Room

Meals of delicious quality and satisfying variety served in a glassed-in dining room at the "Old Farm House."

Fresh sea food of noticeably high quality abounds in the clear waters surrounding Petit Manan Point. Nowhere do fish, clams and lobsters taste better. There are reasons.

Fresh poultry and dairy products, lamb, vegetables, berries, etc., are all raised on the Petit Manan and neighboring farms. Tested cows furnish the milk and cream.

Intriguing trails to tempt the hiker. Fragrant fir balsam, spruce and other groves for those who like to loiter and rest.

The "Old Farm House"

Brochure for Petit Manan in Maine showing the farmhouse, church, a cabin, and the inside of the community building and the dining room that was located on the room to the right side of the farmhouse.

Lafayette and Schoodic National Parks, now called Acadia National Park, Bar Harbor, Frenchman's Bay, Tunk Lake and other scenic and historic places of interest are within easy driving distance of Petit Manan in Maine.

The name "Manan" appears to have been given to the Point, as to the island of Grand Manan, further east, by the early French explorers in honor or remembrance of their home city of Manan in the Department of Maine in Northwestern France.

The Petit Manan Point Road leads off U. S. No. 1 about halfway between Steuben and Millbridge, skirts the base of Pigeon Hill, a sightly eminence just outside the Petit Manan reservation (from which a superb view is afforded of open ocean, picturesque bays, numerous islands and many mountains), passes through the Pigeon Hill Fishing Village, crosses the "Carrying Place" and ascends a short slope to the level of the blueberry fields, thence to the office and headquarters of Petit Manan in Maine at the "Old Farm House," approximately seven and a half miles from U. S. No. 1.

Pleasant trails lead to Goddard Beach, Chair Pond Beach, Big Pond Beach, Bear Cove, Cable Cove, and the Rustic Chapel, Old Club House, Sand Cove Beach and Anchorage in Section I.

The cozy log cabins nestle here and there toward the open ocean.

"One rests from the year at Petit Manan as one rests from the day at night."

A brochure showing the beach and more cabins, it also mentions Lafayette National Park now being named Acadia. The name change took place in 1929, dating this brochure most likely to the early 1930s.

3

Land Company Buildings

Although the different companies were ultimately unsuccessful in creating another Bar Harbor, their efforts did result in several buildings that have historical importance. The clubhouse was turned into a small hotel that housed guests and functioned as a restaurant, and the annex was also used for guest rooms. Because of these facilities, families were able to vacation here and enjoy a more rustic experience that continues to this day. This is a history of the buildings that the land company built, owned, and paid taxes on over the years.The letters and numbers in parentheses identify the location of these buildings on the maps located on pages 66, 67, and 68.

Farmhouse, Barn, and Sheep Shed

S. K. Abbott is on the tax rolls for 1890 with a farmhouse, two barns, and a sheep shed, with two barns listed until 1894.

Hall/Casino (A9)

A hall shows up on the tax rolls in 1886. In 1910, it was called a casino, a designation that it kept until it was crossed off the tax rolls in 1938. The term "casino" at that time was different from what we think of today, it was really a recreation hall by today's standards.

The casino had two bowling alleys, a billiards table, a dance floor, and a large stage with scenery for amateur theater. The pool table and bowling alley on the left of the interior view were eventually sold or donated to the Henry D. Moore Library, whose basement had a separate room for the pool table. The bowling alley is still in the library basement, and the Steuben Historical Society has a goal of refurbishing it. The second bowling alley has been located and is in storage in Cherryfield. It was used for years during the Cherryfield Fair.

An exterior view of the casino.

An interior view of the casino showing the bowling alleys and pool table. The bowling alley to the far left is still housed in the basement of the H. D. Moore Library in Steuben.

Lumber Shed

The lumber shed first shows up on the tax rolls in 1896 under the Petit Manan Land Company. It stayed on the tax rolls until 1909 but is not there in 1910. This building may have been removed or converted into a barn or hog house. No known photos of the lumber shed exist.

Icehouse (A10)

An icehouse first shows up on the 1896 tax rolls under the Petit Manan Land Company, remaining on the books through 1948. An icehouse was located across the road from the Annex and a second icehouse was located on the farm.

When Bill Mague and his family moved into the farmhouse in 1946, commercial electricity was not yet run down to the Point, so they used an icehouse. Electricity was run to the farmhouse in 1948, which explains its disappearance that year.

A view of the icehouse, stable, and water tower across from the Annex.

Remains of the icehouse and stables that were located across from the Annex.

Clubhouse (A8)

The clubhouse first shows up on the 1896 tax rolls and stays on the books up through 1949. The clubhouse has been known by many names over the years, including the Main Lodge and Hotel. In 1923, it was managed by Dora C. Smith.

Above: A view of the clubhouse looking northwest.

Right: A view of the clubhouse looking south.

The fireplace in the clubhouse.

Above: View of the dining room in the clubhouse. Notice the water cooler is marked with the Maine Coast Club name on it.

Right: A view of another dining area at the clubhouse.

Above: A recent view of the foundation of the chimney from the clubhouse.

Left: A recent view of the foundation of the clubhouse.

Above: View of a cistern located next to the clubhouse.

Right: The clubhouse chimney still standing in this photo, probably dating to the 1960s.

The Wharf (A16)

The wharf first shows up on the 1896 tax rolls and remains until 1917. It was probably destroyed by ice during the winter of 1917 or 1918.

Above: View of the steamboat wharf looking south.

Below: View of the steamboat wharf looking north.

A view of the dock house at the end of the steamboat wharf.

The remains of the wharf along the Dyer Bay coastline.

Page Cottage (1)

The Page Cottage on lot 112 shows up on the tax rolls for the Petit Manan Land Company from 1896 to 1899, when it was transferred to Oliver Page. It eventually became known as the Dutton Cottage.

Cottage Lot 105 (A1)

The area of lot 105 contains a foundation that is still visible. It measures approximately 24 feet by 34 feet. A cottage is listed on the tax rolls from 1897 until 1910. The 1904 USGS map shows a structure at about the location of lot 105. The 1911 Steuben tax records show a cottage taxed to E. F. Hanson, but it is not there the following year for E. F. Hanson or John M. Kilgore and the Maine Coast Club. No photo or additional information is known about this cottage.

Atkinson Cottage (A4)

The Atkinsin cottage, spelled with an "I" not an "O," shows up on the tax rolls in 1897 and 1898 but then disappears. No location is given. An article in the *Bangor Daily Whig and Courier*, dated July 13, 1895, states an Isaac C. Atkinson of Portland was among a group visiting Petit Manan. The group included E. F. Hanson and Dr. G. C. Kilgore, so it is possible he was involved in the early development at the point.

Neither the foundation on lot 30 nor the cottage on lot 105 are mentioned in any of the journals kept by Leanora D. Booth or Ted Denniston. It was probably gone early before or shortly after the Dennistons started coming to Petit Manan Point.

Boarding House, Smith House, or Annex (A3)

A boarding house appears on the tax rolls only once, in 1898. In 1899, the Smith House at Petit Manan Point is listed with taxes being paid by Alonzo Smith. Most likely, the boarding house was sold to Alonzo Smith, at which time it became the Smith House. Smith was a resident of Steuben who fished and lumbered to earn a living. The house also had a stable.

Smith purchased lots 38 and 134 in section one. He lived at the point during the summers with his daughter, Helen, and, at times, his son, Byron. After his death, a court case was filed by the heirs fighting over the estate, with testimony taken in Milbridge on March 13, 1914. Byron Smith stated that he was away steamboating in Boston for a number of years. He came home to Steuben in the spring and his father and sister were living at Petit Manan Point. When asked when this was, he stated fourteen years ago, 1899. In other testimony about property his father owned, he said "Oh, yes, a shell of a house at Petit Manan Point."

John M. Kilgore purchased the property back on July 8, 1907. Alonzo Smith held the mortgage that was to last two years on the property. Kilgore did not pay, and Smith's estate sold the mortgage to Liberty National Bank on January 22, 1920. Kilgore finally paid the money to Liberty National Bank on January 2, 1926. The boarding house appears to have been used to house visitors. Leanora Denniston Booth refers to the Smith House as a square box for sleeping quarters. Her uncle, Joe, and his family stayed in the Smith House in 1902 when the Dennistons were staying at the Dutton Cottage. Joseph W. Paul was the president of the Maine Coast Club and the owner of lot 81 in section one. The foundation of the Smith House measures around 48 by 30 feet, so the reference as a square box sounds accurate.

The Smith House is on the tax rolls until 1905. From 1907 on, it is listed with other assets on the Petit Manan tax rolls. John M. Kilgore eventually owned the property, which became known as the Annex. A 1932 note in the tax records says "burned" next to it.

The Annex building with a water tower in the background.

Unknown man at the Annex. Could this be John Kilgore?

Hog House

In 1910, the hog house is first listed. In 1911, five hog houses are listed, but the number falls to two in 1912 and then one from 1913 to 1951 when a note in the tax records says "Gone."

Shop (15c)

A shop first listed 1911 is most likely the shop that is still located on the Mague farm property.

Log Cabin

A log cabin is listed on the tax rolls from 1911 to 1931. In 1930, the Petit Manan in Maine taxes included ten cabins and the log cabin. In 1932, however, the tax rolls list eleven cabins but no log cabin. It may have been included with the other cabins or it may be the cottage sold to Tabitha Fitz-Gerald in 1930 by John M. Kilgore.

A recent view of the machine shop building on the farm.

FISH HOUSE (A18)

On April 26, 1902, the Maine Coast Club and Eastern Fish and Lobster Company signed a lease, allowing the latter to use Pinkham Cove for ten years for $100 a year. Eastern had to build and keep in good repair a substantial dam across the cove near its entrance. The dam had to have a gate that could be easily operated, one wide enough to permit a boat 15 feet wide and drawing 5 feet of water to pass through at high tide. Also, it had to build and maintain a footbridge across the dam.

The lease stated that the Maine Coast Club was a corporation organized by Joseph W. Paul, its president. It also listed John M. Kilgore as the treasurer of the Eastern Fish and Lobster Company.

A fish house owned by John M. Kilgore is first listed on the tax rolls in 1916, remaining until the early 1920s. The fish house was most likely located at the dock next to the lobster dam on Pinkham Cove, which is now known as Over's Cove. The early photos show a small building on the dock at the lobster dam.

Above: View of a boat at the Lobster Pound dam at Over's Cove.

Left: Looking across Pinkham Cove at the Fish House at the end of the dam; the casino is to the left, with Dyer Bay in the background.

Right: Some of the lobsters caught and being taken to the Eastern Fish and Lobster Company at Pinkham Cove.

Below: Eye bolt mounted in the rock on the shore of Over's Cove; this is located at the site of the dam and was most likely used for the dam.

Sawmill

The sawmill shows up on the tax rolls from 1916 until 1927. The first year it was called a mill, then a sawmill, and, at times, a portable mill.

Boathouse

The boathouse shows up on the tax rolls in 1917 and then is not listed again. There were a number of boathouses shown in various photos, so it is unclear why only this one showed up for one year in the tax records. It is possible that it was sold to an individual and then the taxes were paid by that person.

Henhouse

The henhouse shows up on the tax rolls from 1917 until 1920. In 1921, two are listed and two remain on the tax rolls until 1951 when one is shown until 1952.

Pumping Stations (A12, A14, A15)

Two pumping stations appear on the tax rolls starting in 1917 and stay on until 1926 when three pumping stations are listed. The three stations remain on the tax rolls until 1940. Then only one is listed from 1940 to 1953.

Freight/Storehouse

In 1918 and 1919, a freight/storehouse is listed on the tax rolls, but no other information is known on the location. The wharf was off the tax records by 1918, so it seems this may have been a different building that the one at the end of the wharf.

Lookout House (B1)

The lookout house first shows up on the tax records in 1919 and stays there until 1929. It is not shown on the 1930 tax records even though it was still standing after 1930. The caption for a 1980s newspaper photo of the foundation quoted William Mague as saying that it was blown over by a hurricane about fifteen years prior.

A view of one of water towers/pump stations located at Petit Manan Point.

View of the lookout house located on one of the highest places on Petit Manan Point. It was used to show prospective buyers the view and property.

One of the eye bolts that held the guy wires for the Lookout House.

A recent view of the foundation of the Lookout House.

Stables and Garage (A13)

The Gougins stable is first listed on the taxes for the Maine Coast Club starting in 1907 and continuing until 1931. James L. Gougins owned lot 141 section one. In 1897 and 1898, two other stables are listed and then do not show up again. A garage is listed starting in 1925 until 1941.

Powerhouse

The powerhouse first appears on the tax records in 1942 and must have been located near the Mague farmhouse. From 1951 through 1959, it was included in the Mague tax bill on the farmhouse and barn. Petit Manan in Maine Inc. used a direct current (DC) generator to provide power to the cabins, and this was most likely the powerhouse that housed the generators for this.

Remains of a foundation on section one, lot 15, not far from the clubhouse site; the foundation is most likely from one of the stables.

Camp Clam Den and Camp Aerie (A5, A6)

Camps Clam Den and Aerie first show up on the tax records in 1947 and are listed as separate items until 1949. These are the two cabins that were located near the clubhouse but were given names at this time.

Sand Cove Dam (A17)

One of the big selling features that all the old advertising mentioned was the fact that Petit Manan had a sand beach. The Petit Manan Land Company built a dam so that when the water was entrapped by the dam it would warm up to make swimming more enjoyable.

A view of one of the two log cabins owned by the land company.

Above: The second of two cabins that were located near the clubhouse.

Left: The back of the chimney for the cabin.

A view of the dam at Sand Cove looking to the north, with the steamboat wharf in the background.

A view of the dam at Sand Cove looking to the west, with the clubhouse to the left and boathouses along the shore. The larger building above the boathouses is most likely the Annex building.

A view looking south into Sand Cove with the dam.

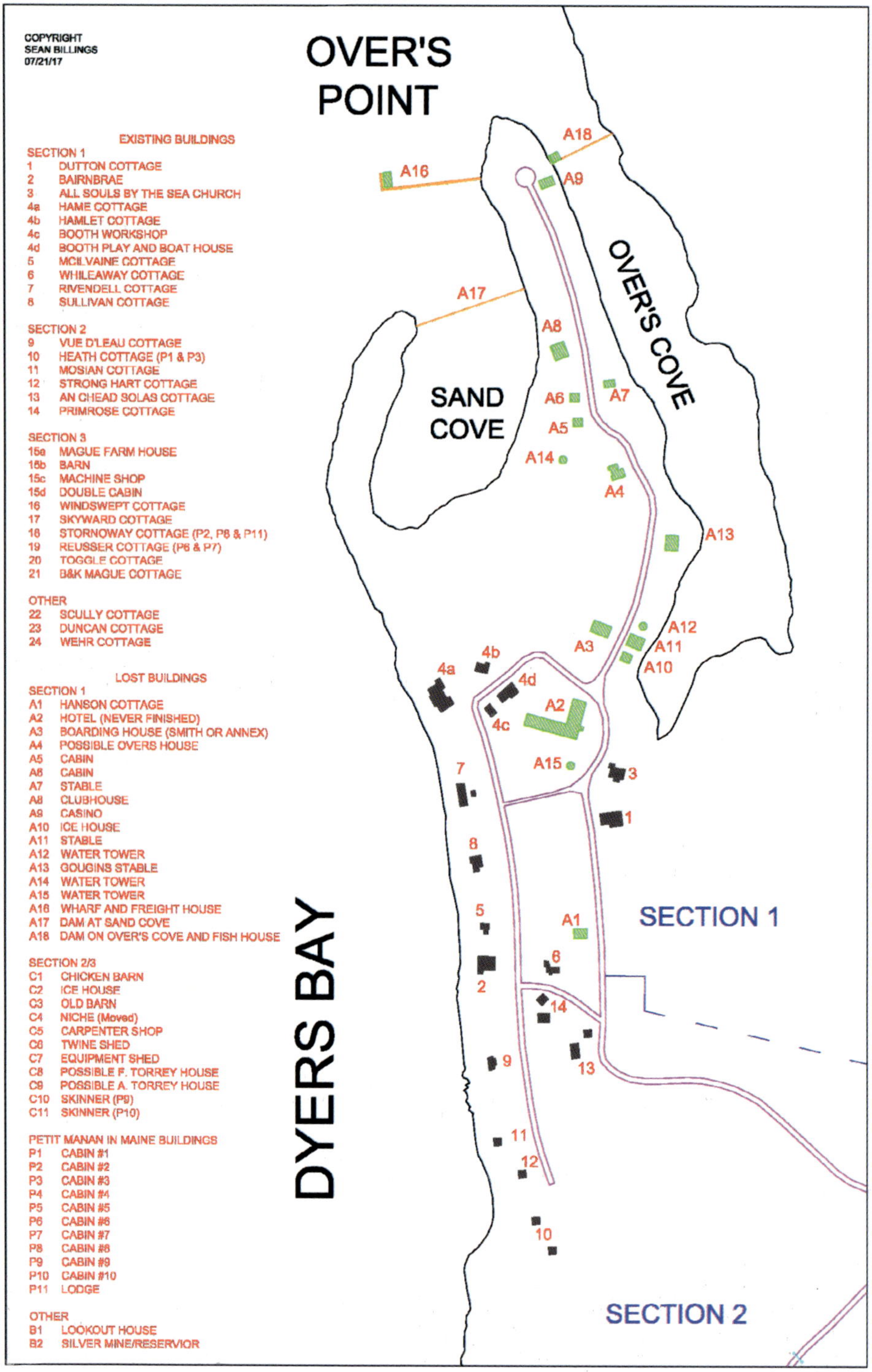

Map showing existing and lost buildings in sections one and two.

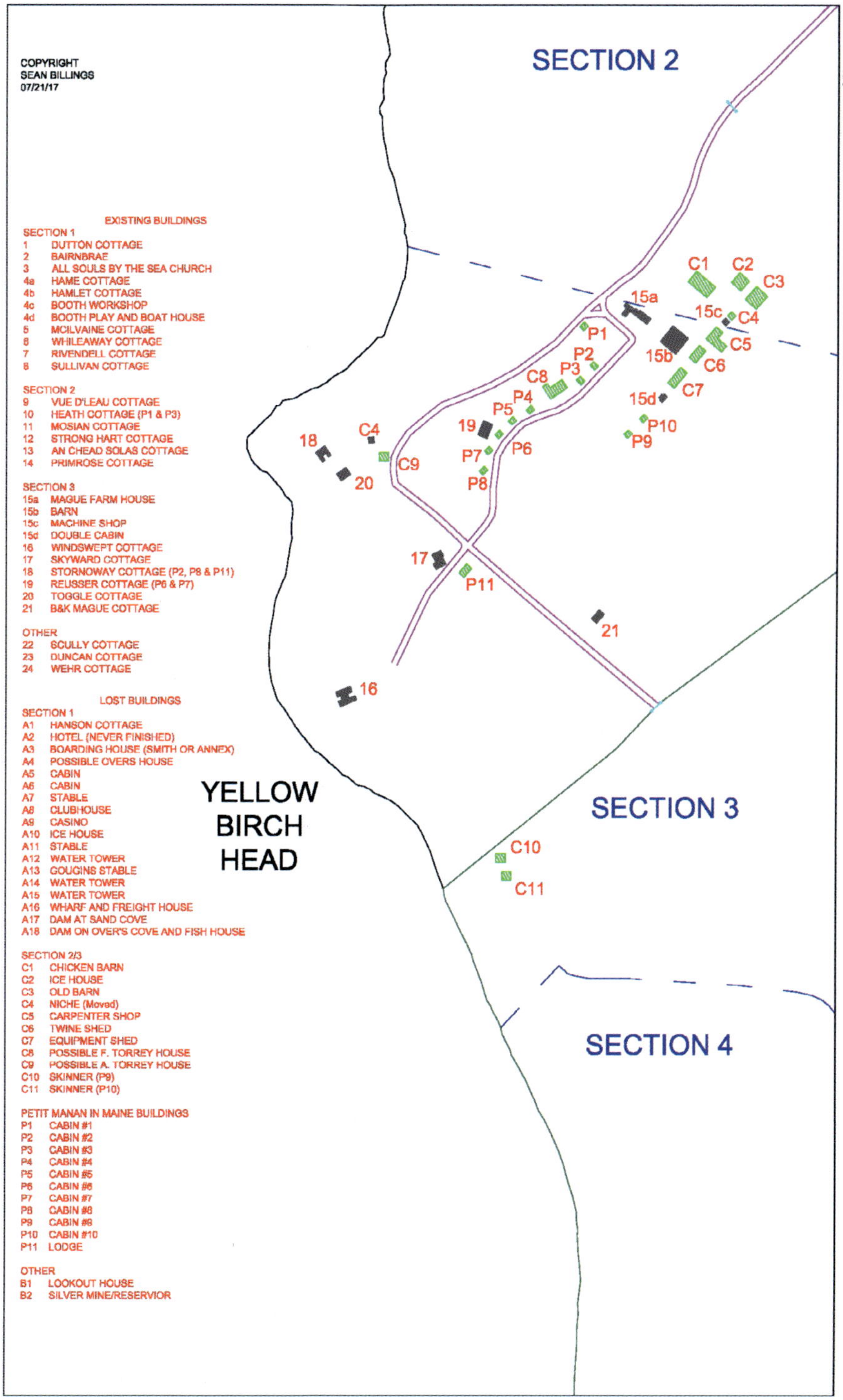

Map showing existing and lost buildings in sections two and three.

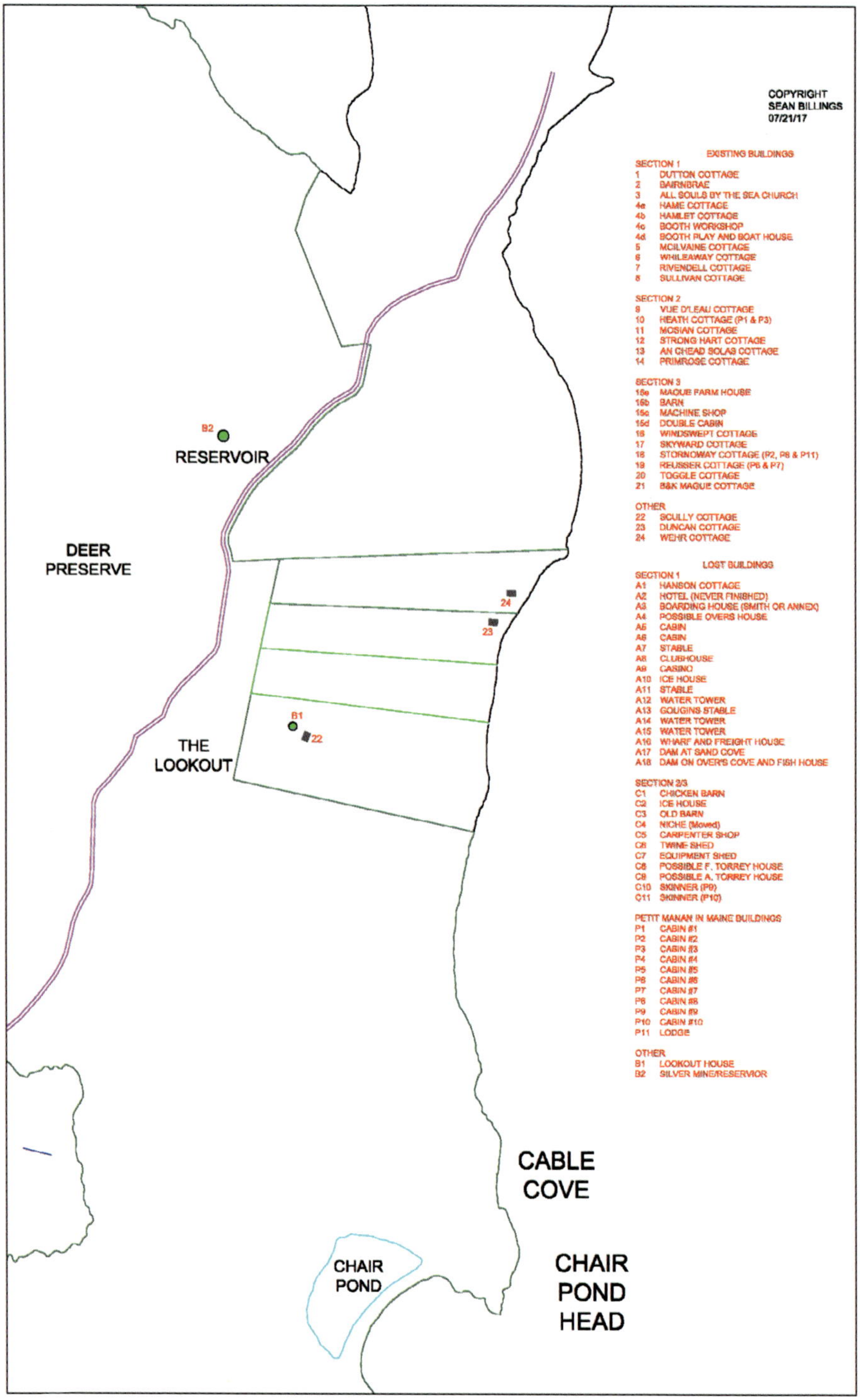

Map showing the existing and lost buildings near Hilton Head.

Modern view of Sand Cove looking into Dyer Bay, remains of the dam can be seen here at low tide.

4

THE COTTAGES

Although development never took hold the way early investors had hoped, their involvement did begin to shape things for several families, many of whom still have ties to the area today. The Denniston family was one of the early families to come to Petit Manan and many of the cottages are still owned by families descended from Eleanor H. Denniston, the Booths, and the McIlvaines. The Magues and Reussers families descend from the Halls and Goddards.

The Petit Manan Land Company and the Petit Manan Land and Industrial Company both spent time and money building the infrastructure at the point. Many of the roads shown on the plans were graded, and the clubhouse, casino, steamboat wharf, and dam for the swimming cove were all built. Lots were sold by both companies, but actual cottages were slow in coming. Both of these companies would fail, and the Maine Coast Club was formed to continue the development.

It appears that the lack of building was of great concern when the Maine Coast Club was formed. Its organization papers and deeds for the individual lots contained clauses to address this. It was specifically spelled out that the grantee "shall construct, or cause to be constructed, on said lot a cottage to cost not less than one thousand dollars, the construction there-of to be commenced within three months from the date of the said deed, and the sum of at least one thousand dollars to be expended thereon within one year of said deed." The deeds also contained a clause that the deed would be null and void and revert to the Maine Coast Club trustee if this did not happen. These deed restrictions seemed to have little effect, and most of the lots continued to be undeveloped even though a number were purchased. Below is a brief history of the cottages that were built.

Section One

Dutton Cottage

One of the first substantial cottages built is what is now known as the Dutton Cottage. Originally known as the Page Cottage, it was constructed on lot 112 in section one. The Page Cottage shows up on town tax records as part of the Petit Manan Land Company starting in 1895 or 1896. It remains until 1900, when the lot was deeded to Oliver L. Page, father of Frederic R. Page, the civil engineer who drew up the plans for the development. Fred R. Page is shown in the 1870 census as living with his parents, Oliver L. and Elizabeth Page, and his brother, Joel Page. Oliver's place of birth is listed as Maine. The 1900 census shows Fred still living with his parents and working as a civil engineer.

The cottage was most likely built by the land company as a model. It resembles drawings that appeared in an early brochure.

Oliver L. Page sold the lot to Gardiner H. Dutton of Steuben on October 15, 1900. The town tax records show Dutton taking over the taxes in 1901. He was the son of William R. H. and Mary J. Shaw Dutton and had two brothers, Henry Shaw and Samuel Robert. His parents were originally from Ellsworth.

An early view of the Dutton Cottage.

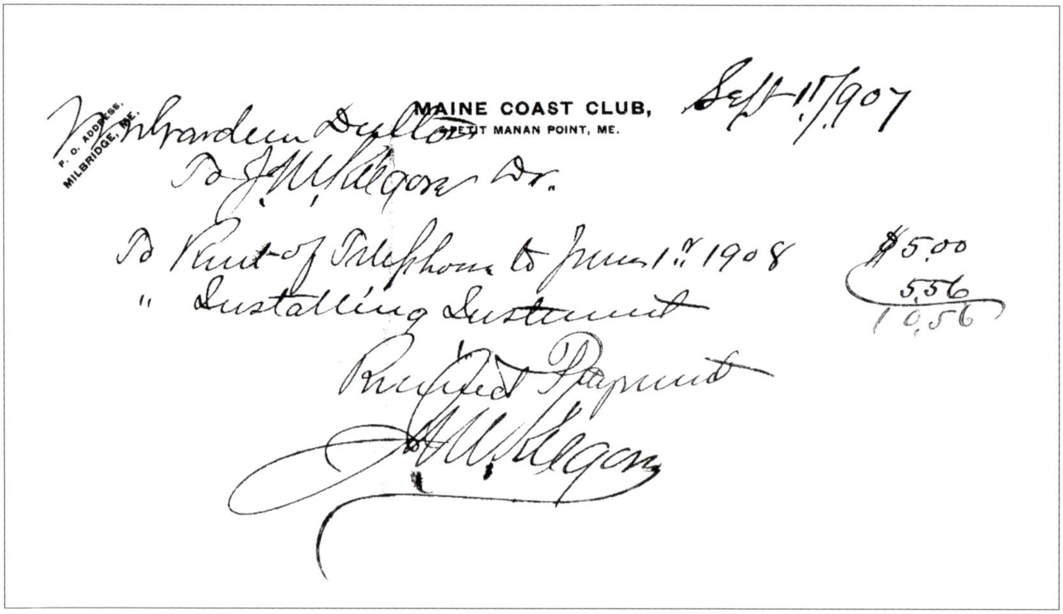

P. O. ADDRESS, MILBRIDGE, ME.

MAINE COAST CLUB,
PETIT MANAN POINT, ME.

Sept 1st 1907

Mr Gardener Dutton
To J. M. Kilgore Dr.

To Rent of Telephone to June 1st 1908 $5.00
" Installing Instrument 5.56
10.56

Received Payment
J. M. Kilgore

The original bill from John Kilgore to Gardiner Dutton from 1907 for the installation of the telephone in the Dutton Cottage.

Recent view of the Dutton Cottage.

Eleanor H. Denniston and her children rented the Dutton Cottage during the summers of 1902 and 1903. The cottage, which had four bedrooms and an attic, had a telephone in 1908, even though it did not get electricity until the 1940s. It originally had a hand water pump in the kitchen sink and a two-hole outhouse off the barn.

On January 23, 1925, John M. Kilgore sold extra land next to lot 112 to Gardiner H. Dutton. On June 2, 1936, the property was transferred to Effie M. Dutton, who owned the property until her death in 1954 when it was passed to her son, Philip S. Dutton. With Philip's death in 1955, the property was passed to his wife, Marion B. Dutton, and his daughter, Roberta Dutton Sawyer. In 1972, Marion died, and Roberta became the sole owner of the property.

On October 9, 2017, Roberta Dutton Sawyer sold the Dutton cottage to James and Donna Hammond, thus ending 117 years of Dutton ownership.

Bairnbrae

In 1904, after spending two summers in the Dutton Cottage, Eleanor H. Denniston decided to build a permanent summer cottage at Petit Manan Point. As a member of the Maine Coast Club, she selected lot 92 in section one. The cottage was designed by a Mr. Weeks, an architect from Bar Harbor. It included ten bedrooms, kitchen, living area, pump house, woodsheds, and garage.

Eleanor was the widow of Philadelphia businessman Edward Evans Denniston who worked at E. W. Clark Company and served as president of the Brooklyn Traction Company. E. W. Clark Company was a banking and financial firm founded in Philadelphia that expanded to New York City, St. Louis, and New Orleans. The firm had such a good reputation that its notes were said to be as good as gold. The U.S. government borrowed money from the firm to finance the Mexican and American War.

Edward E. Denniston died at the young age of thirty-four from typhoid fever, but due to his successful business dealings, his wife was well provided for. He and his wife had five children who survived to adulthood—Ned, Margaret, Anna, Leanora, and Kathleen. Eleanor and her children spent summers at Bairnbrae, beginning a tradition that continues with her descendants today.

Photo of Eleanor H. Denniston, one of the directors of the Maine Coast Club, founders of All Souls by the Sea Church, and guiding forces behind Petit Manan Point in the early years.

View of the porch facing the bay at Bairnbrae.

A recent view of Bairnbrae. (*Photo by Mike Denniston*)

HAME

The HAME Cottage is a result of the tradition begun by Eleanor Denniston. In 1914, Eleanor's daughter, Leanora, married Henry Booth and began coming to Bairnbrae on vacations. After starting a family, they purchased lot 81 in section one from Margaret Paul, wife of Joseph W. Paul, who was once president of the Maine Coast Club. This would have been the lot that Joseph W. Paul received upon becoming a director and member of the club. No record has been found to show that the Pauls ever built a cottage. Personal journals often mention that they stayed at various locations such as the Smith House or Dutton Cottage.

In 1929 and 1930, the Booths built their cottage, which they named HAME using the first letter of the names of their four children—Harry, Amy, Mary, and Eleanor. Jim Strout of Milbridge built the cottage and Cora Duniver, who lived in the farmhouse at the Point, built the chimney and dug the well. The Booths purchased other lots over the years so they would have more space near their cottage. In 1949, the Booths added another cottage called Hamlet to accommodate the growing families of their children. Shortly thereafter, in 1950, a combination of workshop, playroom, and boathouse was built by Ivory Parritt. In 1952, a swimming pool was added and in use until at least 1962. George Marshall built it with help from Jack Strout, whose horses cleared the site. In 1956, tennis courts were added and, in 1958, a wood shop called Carpenters Hall. The cottages are still owned by the Booth family and their descendants.

An early view of HAME.

View of HAMLET after it was first delivered.

A recent view of HAME.

A recent view of HAMLET.

A recent view of the studio at HAME.

A recent view of carpenter's hall at HAME.

McIlvaine Cottage

The McIlvaine Cottage is located on lot 91, next to Bairnbrae. The log cabin is listed on the Steuben tax records from 1911 to 1931, with ownership moving from Kilgore to Shumaker to Petit Manan in Maine. The log cabin is no longer shown on the tax records for Kilgore or Petit Manan in Maine Inc. after 1931, however. This coincides with lot 91 being sold by John M. Kilgore to Tabitha Fitzgerald of Boston on April 9, 1930.

Fitzgerald was born on December 17, 1875 in Boston to Thomas and Martha Fitzgerald. Both Thomas and Martha were born in England, and Thomas was a tailor. The 1910 Boston census shows the Fitzgerald daughters, Tabitha, Elizabeth, and Mary, living with their parents. Thomas is listed as retired; Tabitha and Mary were teachers. In 1920, Tabitha is listed as the head of household with Gertrude Smith a boarder. Tabitha continued to be listed as a teacher. The 1930 census shows Tabitha living alone and working as a teacher.

The Steuben tax records for 1930 show Tabitha paying tax on a cottage. She sold the property to Helen D. Perkins on August 31, 1946. On September 7, 1949, Perkins purchased the adjacent lot, number 90, from Petit Manan in Maine.

On December 7, 1964, Perkins gave the cottage and property to Amy Booth McIlvaine, whose family still owns it.

A recent view of the McIlvaine Cottage.

Whileaway Cottage

Whileaway is located on lots 93 and 106 in section one. In 1902, the Maine Coast Club sold the lots to W. C. Sperl of Newton, Mass. His wife, Martha L. Sperl, was widowed in 1914, and the Town of Steuben took the property for taxes around 1922. On July 15, 1924, T. Dudley Drew-Bear of Philadelphia bought the property from the Town of Steuben. On April 4, 1925, Martha L. Sperl transferred her ownership Drew-Bear, and on September 15, 1925, a Kenneth K. Sperl transferred his rights to the property to Drew-Bear.

Dudley Drew-Bear, who was born in England in 1897, shows up on the 1920 census in Philadelphia as a mechanical engineer living with his mother, Jessie; brother, Robert; and sister, Joy. Jessie is listed as a florist and head of the household.

The cottage first shows up on tax records in 1928. Up until that time, the tax leverage for the property had been based on an empty lot. Shortly after buying the property, Drew-Bear moved to Venezuela and married. He transferred ownership of the cottage to his mother, Jessie H. Drew-Bear of Philadelphia, on May 31, 1929. A brochure for Penzance-in-Maine quotes her as saying: "Since building our cottage at Petit Manan, it has become our real home. The entire family hurries to Maine at the approach of spring and is always loath to heed the city's call when autumn arrives."

Jessie H. Drew-Bear sold the cottage to William Brewster Taber on October 24, 1933. Taber owned the cottage for only a few years before he sold it to E. Roberta Puffer, a

A 1937 view of Whileaway Cottage.

A recent view of Whileaway Cottage.

teacher who taught in New Jersey and Massachusetts. Puffer, who had been born in Columbia, Maine, deeded the property to her first cousin, Irene Puffer Morrison, and her husband, John Morrison, on June 26, 1951. John Morrison was a congregational minister who spent time with his wife and their four children at Petit Manan during July or August each year. He would sometimes lead the worship services at All Souls by the Sea. When he died in 1969, Irene P. Morrison became the sole owner. On December 9, 1976, she transferred ownership to her four children, who are currently all co-owners.

Rivendell

Rivendell is located on lots that were all originally owned by separate people. The Petit Manan Land Company sold lot 87 to "Badger" and "Manley," whose full names are unknown, on July 16, 1894. Horance N. Hastings bought lot 85 from the Petit Manan Land Company. Marianne Whitney, Ella H. Amory, and J. Frank Turner bought lots 86, 88, and 89 respectively from Maine Coast Club.

On March 7, 1901, Badger and Manley sold lot 87 to Lillian E. Cassens. She was the daughter of Edgar Hanson and wife of R. Herman Cassens, owner of the Eastern Illustrating & Publishing, a company that took photos at Petit Manan Point and made postcards out of them. Lillian sold the lot to Ella Howard Amory on November 3, 1902. The first time a cottage shows up on the property according to Steuben tax records is in 1911. That year and the following year, a log cabin is listed with the taxes being paid by J. M. Amory. On October 16, 1925, the Estate of Ella Howard Amory sold the property to Frank T. Whitney.

The 1923 to 1926 tax records show J. M. Amory owning both lots 87 and 88, and F. T. Whitney owning only lot 86 with no house on it. The 1927 tax records are the first that show a house on lot 86. The 1928 tax record contains a note that lots 87 and 88, previously owned by J. M. Amory, are now owned by F. T. Whitney and include an added garage. The 1929 tax records show F. T. Whitney owning lots 86, 87, and 88 with a house and garage. On May 23, 1939, the property is transferred to Frank T. and Alexandra Dushane Whitney. The Whitneys sold the property to Margaret Speaks Pearl on May 26, 1947.

On May 25, 1950, Margaret Speaks Pearl purchased lots 85 and 89 from Carmen G. Whitten, Maurice H. Whitten, and John F. Whitten. The Petit Manan Land Company sold lot 85 to Horance N. Hastings and then Mahlon A. Whitten sold it to Hastings & Sons Publishing on August 15, 1924. Whitten also purchased lot 89 from the Town of Steuben on July 8, 1924.

Margaret Speaks Pearl sold the property to Edwin H. Craig and Margaret Smythe on September 1, 1955. Edwin H. Craig made Margaret Smythe the sole owner on July 31, 1965. She sold the property to Robert G. and Georgiana M. Feeney on March 20, 1971. The property is still in the Feeney family. The Feeneys sold lots 87 and 88 and part of 89 in 1996 as detailed below.

A recent view of Rivendell Cottage.

POTO (Place On The Ocean) Sullivan Cottage

On August 5, 1996, Charles A. and Dawna P. Sullivan purchased lots 87, 88, and part of 89 from Robert G. and Frances M. Feeney. The Steuben tax records show that the Sullivans built the cottage in 2000. The property it is still owned by the Sullivan family.

SECTION TWO

Vue d'leau

On May 16, 1949, Agnes E. Dardis purchased a piece of land from Eleanor G. Daniels. The land was located at the very far north end of section two on the border of section one. On the original plans, this area was where Forest Park and Vue De L'eau drive came together. Forest Park was a circle that was originally shown as open space on the land plans. Agnes's son, Thomas D., built a cottage on the land near the water starting in the summer of 1950. The framed sections were left covered for the winter, and the piers and framing were erected, and rough clapboard siding put on in 1951. On March 8, 1950, Agnes bought an additional adjacent piece of land from William R. Mague.

The cottage first appears on the Steuben tax records in 1954. It was originally one room but was expanded to include larger sleeping quarters. The south wing was added in 1956 and the north wing in 1959. Electricity and plumbing were added to the cottage in 1969.

The Dardis cottage, Vue d'Leau, when it was first built.

A recent view of Vue d'Leau Cottage.

The land the cottage is on was transferred from Agnes E. Dardis to Thomas D. and Mary C. Dardis, and they purchased an additional tract of land from William R. Mague. On October 19, 1984, Thomas D. and Mary C. Dardis transferred the ownership to the Dardis Corporation, which transferred ownership to the Dardis Family Trust on June 28, 2005. The Dardis family still owns the cottage.

An Chead Solas

Peter Dardis, son of Thomas and Mary, purchased property from Mague Inc. on November 12, 1986. His cottage shows up on tax records as being built it 2003

Primrose Cottage

James D. and Margaret D. Kovacs exchanged lots 3 and 4 near the end of Over's Point for a lot across from the Whileaway Cottage on lots 93 and 106 on September 15, 2010. Margaret is the daughter of Thomas and Mary Dardis. The Kovacs built a cottage on the property, which shows up on tax records as built in 2010.

Moisan

The Mosian Cottage is located near where lots 240 and 241 are shown on the plan in section two. On March 9, 1950, William Mague sold the land to Stuart E. Phelps. It was then transferred to Stuart E. Phelps, Jr., on May 22, 1967. He sold the property to Cheryl M. and Richard L. Moisan on May 28, 1981. The Moisan cottage first appears on the Steuben tax records in 1987.

Strong Hart (Suzanne Armstrong)

The Armstrong Cottage is located near lots 242 and 243 as shown on the plan for section two. The land was part of the property sold by William Mague to Stuart E. Phelps on March 9, 1950. Stuart E. Phelps sold this southern portion to Bruce R. Phelps on May 5, 1967. Bruce sold to Walter R. and Suzanne Armstrong on December 15, 1980. The Armstrong cottage first appears on the Steuben tax records in 1992. Suzanne Armstrong has owned the property since November 15, 1995.

Heath

The Heath Camp is located in an area of section two that was originally shown as open space. Douglas and Harriet E. Heath purchased the first of three lots from William R. Mague on September 14, 1953.

The 1954 Steuben tax records show Harriet paying taxes on the land only. In the 1955 tax records, a note was added to her listing that says, "cottage of Bill Mague." Two cabins, one regular sized and one large, were purchased from Petit Manan in Maine. The 1956 taxes records show it as being a log cottage. The Heaths purchased cabins number 1 and 3 from Petit Manan in Maine to make the camp. Bill Mague then transferred a second adjoining lot on August 2, 1966 and a third adjoining lot on September 12, 1968.

The Heaths, who were originally from Philadelphia, kept the two cabins separate so the smaller one could be used by visiting friends. The Heaths used the larger cabin, to which a kitchen and extra bedrooms were added over the years. A stone fireplace was added by Douglas and his father in law, Glenn Frye. At the time of this writing, the stone fireplace was being removed by Russell Heath due to deterioration. The cabins are still owned by the Heath family.

The small Heath Cottage previously at Petit Manan in Maine.

The larger Heath Cottage before the fireplace was removed, previously a Petit Manan in Maine cabin.

Section Three

The Farmhouse

Determining the date that the existing farmhouse at Petit Manan Point was built is difficult. The 1865 tax records show Albion Hinkley paying tax on a house and barn and 200 sheep on Petit Manan Point. The 1871–75 tax records show Nathan Hinkley and Albion Hinkley paying taxes on half house each and 400 acres each at Petit Manan Point. During this same timeframe, Henry Over had a 100-acre farm on Petit Manan Point and Edward Hilton had 47 acres with a house and barn near Carrying Place Cove. From 1876 to 1881, the Petit Manan Sheep Company was taxed on two houses, one barn, and one sheep shed. In 1884, Sylvester K. Abbott was being taxed on half a new house and half an old house at the Point; the other halves of the houses were being paid by Charles J. Smith. In 1884, Abbott paid taxes on 800 acres and the 100-acre Over place. Smith paid on 800 acres. In 1885, Abbott is paying taxes on 1,600 acres, the 100-acre Over place, 47-acre Hilton place, two barns, one house, and one sheep shed. The Overs lived at the Point since 1850 when they bought the land from Wilson M. Dyer. If there were two houses in 1884, the Over house was probably the older one and farmhouse was the newer one. Did the Petit Manan Sheep Company build the "new" house in 1876? It would stand to reason that the Over and Hilton houses were removed around 1885.

During the time the Point was being developed, a number of people lived in the farmhouse and raised food for the people living there and visiting. Jeremiah and Susan Grant are mentioned as living in the farmhouse in the All Souls by the Sea booklet by Glenore Orris Coblens. In the 1900 census, Jeremiah and Susan Grant are listed as living in Steuben, and Jeremiah is listed as being the superintendent for a land company. By 1910, the Grants are no longer in Steuben, according to the census.

John and Cora Duniver are referred to in several journals as living in the farmhouse, supplying food, and doing odd jobs for the summer visitors. Cora L. Duniver, who was born on December 14, 1872, and John S. Duniver, who was born on March 21, 1850, show up in the 1900 Steuben census. John's occupation is listed as fisherman. In 1910, the couple appears again, with John's occupation listed as teamster for farm work and Cora as a farmer.

John Duniver died on September17, 1925. In the 1930 census, Cora is again listed as a farmer. Living with her was her sister, Abbie Leighton, and granddaughter, Eugenie E. Davis. Cora died on March 6, 1940.

An advertising pamphlet put out by John M. Kilgore further confirms that the area was indeed being actively farmed. The pamphlet states that "management maintains an up-to-date farming proposition, it enables them to supply the tables of our lodges, camps and summer residences on the property with fresh vegetables, milk, cream, butter, eggs, all pork products, lamb, mutton, veal and chicken, all produced on the property."

On April 16, 1940, Eleanor G. Daniels purchased the property containing approximately 2,000 acres from the Mechanics National Bank of Worcester, which had foreclosed on the property in 1929. The bank had acquired the mortgage John M. Kilgore had on it from Charles A. and Frank H. Steward.

Daniels was a shareholder in Petit Manan in Maine Inc., a vacation community operated by Winthrop G. and Madeline T. Hall. Petit Manan in Maine Inc. brochures list the farmhouse differently. In one, the photo is captioned as he "Old Farmhouse" and in another it is captioned as the "Tavern."

A *circa* 1930 view of the farmhouse at Petit Manan Point.

Haying at the farm, with Mary, Harry, and Eleanor Booth in the hay wagon.

A modern view of the farmhouse.

In 1946, William and Priscilla Mague moved into the farmhouse year-round and began growing blueberries and weir fishing. On March 6, 1950, Daniels sold the farmhouse and 150 acres around it to William R. Mague, husband of one of the Halls' daughters, Priscilla.

The "Double"

The "Double" is located on the Mague farm and is one of the original Petit Manan in Maine cabins. It was built after the first ten single-bedroom cabins and got its name because it had two bedrooms.

Reusser Cottage

William R. Mague transferred land to Verdi L. and Margaret Hall Reusser on July 28, 1954. The Reusser cottage was made from two of the Petit Manan in Maine cabins, numbers 6 and 7. The timeframe during which the two cabins were used for the Reusser cottage coincides with two cabins being removed from the Petit Manan in Maine Inc. tax rolls in 1953. The Reusser family still owns the cottage.

A recent view of the Double Cottage from Petit Manan in Maine.

The Reusser Cottage, previously cabins from Petit Manan in Maine.

Skyward

The Skyward cottage is located on lots 113–120 in section three. These lots were part of the property foreclosed on my Mechanics National Bank of Worcester. Petit Manan in Maine Inc. purchased the lots from the bank and then resold the lots to Robert Simpson on December 4, 1930. In 1941, the cottage was built. Robert Simpson died on March 10, 1950, and Eleanor G. Daniels purchased the property from his heirs on November 1, 1950. The property was purchased on January 20, 1948 for back taxes by David H. and Spencer Leighton. To clear up the ownership on February 4, 1959, Daniels bought the property from Leighton.

On November 27, 1970, Daniels sold the property to Robert M. Mitchell. On March 4, 1971, it was transferred to Robert M. and Mary Ann Mitchell. On October 24, 1985, it was transferred to Frances C. Mague and, on November 5, 1985, it was transferred to Frances C. Mague and Glenore O. Coblens. On June 30, 2010, it was transferred to the Frances C. Mague Trust.

Stornoway

On May 5, 1943, Eleanor G. Daniels sold a 3-acre piece of land on the northern tip of Yellow Birch Head to Milton T. and Marjorie Hall Stauffer. The original plan showed this as open space below Vue De L'Eau Drive. The Stauffers built a camp on the site using cabins from Petit Manan in Maine. The cottage first shows up on the tax records in 1950. Notes on the tax records call it out as two cabins and a community house. The cottage was made up of the Petit Manan in Maine community building and cabins 2 and 8. The 1950 date also coincides with the time that two cabins and the community house were removed from the tax levy of Petit Manan in Maine Inc.

The Stauffers had three sons who were pilots in the military during World War II. One of them was stationed at Stornoway, on the Isle of Lewis in Scotland. He told his mother that the coast at Stornoway reminded him of the coast at Petit Manan, and so the family named the cottage Stornoway. The Stauffers transferred the property to Richard C. Coblens on June 16, 1960. The Richard C. Coblens Estate transferred the property to Glenore O. Coblens on November 21, 1984 and it was transferred to Frances C. Mague on July 24, 1989.

Toggle

The cottage Toggle was built in 2012 by Jessica Mague, Frances Mague's daughter. The cottage is located on the same lot as Stornoway. It measures 12 feet by 18 feet and is the small child to the larger Stornoway. The name Toggle refers to the small float that accompanies the larger lobster floats on some lines.

Windswept

On January 28, 1912, the Maine Coast Club sold lots 121, 122, 123, and 124 to G. Horton Glover. In 1927, Glover also bought Eastern Island off the tip of Dyer Point. On July 9, 1928, Glover bought lot 125 from John M. Kilgore. The cottage first shows up on the tax rolls in 1928 along with a garage. The 1929 tax book shows Glover paying taxes on the house, garage, pumphouse, boathouse, two sheds, and a wooden tennis court. The pumphouse was possibly used for a concrete swimming pool that was located on the property on the bluffs overlooking the ocean. Folklore

A recent view of Skyward Cottage.

A modern view of Stornoway.

indicates that Glover had a daughter with polio, and he built the pool so she could swim.

On July 22, 1936, the property was transferred from Glover to Milton T. Stauffer. On August 11, 1937, Milton Stauffer transferred the property and cottage to Eleanor G. Daniels.

Mary Ellen Chase rented the cottage in 1940 from Eleanor G. Daniels. Born in Blue Hill on February 24, 1887, Mary Ellen Chase graduated from the University of Maine and held master's and doctorate degrees from the University of Minnesota. She taught at the University of Minnesota for eight years and then moved to Smith College in Massachusetts. She would eventually write sixteen novels, as well as fourteen volumes of essays. Inspiration for her most famous novel, *Windswept*, came while she was renting the cottage in 1940. The novel came out in 1941 and became a bestseller in 1942. With the proceeds from the book, Chase purchased the property and cottage from Eleanor G. Daniels on December 18, 1941. Chase named the cottage Windswept, a name that continues to be used today.

Chase transferred the property to Leslie S. Pearl on August 25, 1955. On September 29, 1965, Leslie S. Pearl transferred the property to Jane A. Skinner, who transferred it to Peter C. Huycke and Jane S. Huycke on January 13, 1989. On October 19, 2007, the property was transferred to Peter C. Huycke and Lynda Gentile-Huycke, who transferred the property to Mark V. Tremallo on January 5, 2010. On October 4, 2012, Tremallo transferred half ownership to Wellner A. Poppe. The property was listed on the National Register of Historic Places in 2007 because of its relationship to Chase. After purchasing Windswept, Mark and Wellner Tremallo had the cottage redone, replacing wooden posts serving as a foundation with a basement.

A recent view of Windswept, named by Mary Ellen Chase after her famous novel, *Windswept*.

The remains of a pool at Windswept.

Ben and Kathy Mague

Ben and Kathy Mague built a cottage in 2012 on land near the end of the Mague property on the road going out to Bear Cove.

Scully Cottage

In the mid-1960s, John J. and Margorie M. Scully were looking for a place to relax on the Maine coast. John was getting ready to retire from the Portsmouth Navy Yard in New Hampshire. On a whim, they drove down Pigeon Hill Road and ended up at Bill Mague's house on the Point. They struck up a friendship, corresponding over the course of a few years until Mague offered to sell them a piece of property.

The Scullys purchased around 30 acres at the site of the old lookout house. The property was transferred from Petit Manan Industries to John J. and Marjorie M. Scully on March 22, 1967. The Scullys built a cottage near the location of the former lookout house, preserving the lookout house foundation and having a scale replica made of it.

The cottage is now owned by the Scullys' son, George.

A model of the Lookout House the Scullys had built.

A recent view of the Scully Cottage located near the site of the Lookout House.

Duncan Cottage

The lot north of the Scullys' was sold to Donald B. and Janet M. Duncan by Petit Manan Industries on October 13, 1969. The lot contained approximately 25 acres. The Duncans built a cottage near the water. After their death, the property went to their daughter, Julia Duncan Rinehart, who subdivided the property into two approximately 12-acre lots. The lot with the Duncan Cottage was sold on August 26, 1996 to Michael and Gertrude M. Sloan, who sold it to the Barbara W. Parson Trust on June 4, 2008. After Barbara's death, it went to Wendi A. Parson and Kent R. Parson on July 21, 2015 and then to Kent R. Parson on February 17, 2016.

On December 15, 2017, Kent Parsons sold the cottage and lot to the United States of America, and it became part of the wildlife refuge.

Wehr Cottage

The next property to the north, next to the Duncan property, was purchased by Paul Wehr from Petit Manan Industries on January 12, 1972. The lot contained approximately 13.23 acres. The Wehrs built a cottage down near the water and the property remained with the Wehrs until December 8, 2015, when they transferred it to the Friends of the Maine Islands National Wildlife Refuge. The parcel was transferred to the United States of America on June 30, 2018 and is now part of the wildlife refuge.

A recent view of the small Wehr Cottage.

Opposite above: A recent view of the Duncan Cottage, now named Cedar Lodge by the refuge and used as living quarters for interns, volunteers, employees, and visiting researchers.

Opposite below: A recent view of the Wehr Cottage.

5

ALL SOULS BY THE SEA

With the closest church being in Milbridge or Steuben and the mode of travel in 1906 being horse and buggy, Eleanor Denniston decided she wanted a church on Petit Manan Point. Edmund Gilchrist, a student and family friend of the Dennistons, had spent time at Petit Manan Point over the years as their guest. He designed the church and supervised its construction. Philadelphia Architects and Buildings' website shows it being built in 1910. The church was Gilchrist's first building design in Maine and only the second design of his professional career.

Edmund Beaman Gilchrist was born in 1885 in the Germantown neighborhood of Philadelphia. He spent one year at the Drexel Institute and two years at the University of Pennsylvania before starting to work, one year with Horace Trumbauer and five years for Wilson Eyre. In 1911, he started his own firm. He also designed the Ellsworth Town Hall, Islesford Museum, Abbe Museum in Acadia National Park, and hundreds of houses and businesses in Philadelphia and Maine. He retired in 1943 and lived until 1953.

All Souls by the Sea was built on lot 130 in section one, next to the Dutton cottage. The land was officially donated by John M. Kilgore on January 6, 1914 to trustees John M. Kilgore, Eleanor Denniston, and Lester V. Bailey.

John Kilgore donated the pump organ. The bell in the tower is said to have come from an old ship, according to Leanora D. Booth's *Petit Manan Journal*.

The first minister to hold a service was Dr. George Foley of Philadelphia in August 1911. Allen R. Van Meter held services in 1914–15 and also in 1926–27.

The following is a summary of the events that were documented in the All Souls by The Sea Church minutes.

During an August 4, 1957 meeting, it was determined that a number of the residents were interested in a small burial ground adjacent to the church. Bill Mague made it known during the meeting that land would be available to the church if the Sea Scouts of Maine returned the land that they possessed in section one, which he understood they were contemplating. During an August 11, 1957 meeting, a committee made up of Henry D. Booth, Jr., Edward E. Denniston, William Mague, Winthrop Hall, and Milton Stauffer recommended that the church proceed with the act of incorporation.

During the first meeting held on August 11, 1957, the following officers were elected: Henry D. Booth, Jr., chairman; William R. Mague, vice chairman; Anna H. Denniston, secretary-treasurer; and Edward E. Denniston, clerk. Additional board members were Margaret Smythe, Milton T. Stauffer, and Edward E. Denniston, Jr. Also, during this meeting, the by-laws were read and adopted. The church was incorporated under the laws of Maine on October 3, 1957.

During the August 9, 1958 meeting, it was noted that burial grounds had been obtained by the church. The officers of the church were re-elected. The burial ground dedication was held August 17, 1958.

During the August 30, 1959 meeting, Henry Booth was elected chairman; Anna Denniston, treasurer; and William Mague, clerk. The office of secretary left vacant by Edward E. "Ned" Denniston was filled by Margaret Smythe. Henry Booth informed the group that he had obtained the Sea Scout property and wished to deed it over to the church.

During the August 7, 1960 meeting, it was reported that the former Sea Scout property had been transferred to the church. A property committee consisting of Leslie Pearl, Ed Craig, and Tom Dardis was appointed to maintain the land.

The August 13, 1961 meeting included a discussion on removing the chimney from the old clubhouse. Hall stated that after cleaning off some old bricks, he did not think the chimney would fall down, so it was moved to keep it standing as a landmark. The officers elected at the meeting were Henry Booth, chairman; Margaret Smythe, secretary; Anna Denniston, treasurer; and Harold Handy, clerk.

For 1962 and 1963, the following people served on the board: Henry D. Booth, Jr., president; Edward E. Denniston, vice-president; Margaret Smythe, secretary; Anna Denniston, treasurer; and Harold Handy, clerk.

In 1963, the church made donations to Sunbeam Maine Coast Mission and Eastern Memorial Hospital in Ellsworth.

In 1964, Winthrop Hall was added to the property committee and Edna R. Kindell was added as assistant treasurer to the re-elected board. A donation was made to Maine Coast Memorial Hospital in Ellsworth.

At the August 15, 1965 meeting, a discussion was held because the tax bill was received for all land in the church name other than the land the church sat on. Edward Evans Denniston, Jr., volunteered to take the land with his family and pay the taxes, preserving the land in its present condition. A motion was made for the church to retain ownership of the dock and pier and assume the taxes on it. Leslie Pearl and Ed Craig were no longer property owners, so two vacancies existed on the property committee. Richard Coblens and Alfred L. Byron were elected to fill those vacancies. Henry D. Booth, Jr., asked not to be re-nominated as president since he had already served eight years. Edward E. Denniston, Jr., was elected as president and Winthrop G. Hall was elected vice-president. All other officers remained the same.

During the August 14, 1966 meeting, it was reported that the property the church had been being taxed on was transferred to Edward Evans Denniston, Jr., and Henry D. Booth III, and that they would share the responsibility of paying the taxes. The elected officers remained the same, except Leanora D. Booth, who was elected as secretary, replacing Anna Denniston.

In 1967, the Petit Manan Boat Club was organized to maintain the dock that the church owned. Bill Mague donated a 40-foot-wide by 75-yard piece of land behind the church.

At the August 13 meeting, the officers elected were the same as the previous year. Wesley Harrington was elected to the property committee to take Winthrop Hall's place.

At the 1968 meeting, the officers were re-elected. Spencer Leighton was added to the property committee. A donation was made to the Maine Seacoast Mission.

At the August 1, 1969 meeting, the board decided to donate again to the Maine Seacoast Mission. The officers elected were Edward E. Denniston, president; Winthrop Hall, vice-president; Leanora D. Booth, treasurer; Amy McIlvaine, secretary; and Harold Handy, clerk.

The August 9, 1970 meeting included a discussion on donating to the local community. It was decided to donate to the Maine Seacoast Mission and to the Milbridge Medical Center. The same slate of officers was elected with the addition of Gay Coblens as assistant treasurer.

During the August 22, 1971 meeting, the same slate of officers was elected as in the previous year. The church donated to the Steuben Fire Department, the Maine Seacoast Mission, and the Milbridge Medical Center. The decision was made to open the church to anyone throughout the year for weddings or other religious events. Handy Pinkham or Priscilla Mague would handle the church events in the months other than July and August.

A meeting was held on August 13, 1972, when Gay Coblens was commended for her work on the church organ. Donations were made to the Steuben Fire Department, Maine Seacoast Missionary Society, and the Narraguagus Nursery School. The first booklet on the history of the church was printed and the officers are re-elected.

Clerk Harold Handy reported at the August 12, 1973 meeting that a wedding had taken place the previous fall. Bill Mague reported that beginning the following year, the Mague dump would be closed. He would be bulldozing it over in accordance with current landfill practices. Donations were made to the Steuben Fire Department and the Narraguagus Nursery School. The same slate of officers was re-elected.

During the August 11, 1974 meeting, donations were made to Narraguagus Nursery School and the Steuben Fire Department. Bill Mague discussed his efforts with the government to preserve the point. His efforts received a hearty approval and prayers of all members of the All Souls community. Leanora Booth retired as treasurer, Gay Coblens replaced her, and Marge Scully became assistant treasurer. All other officers remained the same.

At the August 17, 1975 meeting, Bill Mague reported that nothing had been finalized yet with his work on the wildlife refuge. Donations were made to the Steuben Library, Steuben Ambulance, Steuben Fire Department, and the Narraguagus Nursery School. Winthrop Hall retired as vice-president and Bill Mague was elected to be vice-president in his place. All other officers remained the same.

During the August 15, 1976 meeting, the same officers were elected. Bill Mague discussed the wildlife refuge's regulations and asked that no one drive beyond his house when going to Bear Cove. Donations were made to the Narraguagus Nursery School, Steuben Fire Department, Petit Manan Ambulance Company, and the Steuben Library.

During the August 14, 1977 meeting, it was reported that the Scullys', Wehrs', and Duncans' mailboxes had been vandalized. Donations were made to the Steuben Library, Steuben Fire Company, and the Petit Manan Ambulance. The slate of officers was re-elected. Tom Dardis resigned as the property committee chairman and as the commodore of the boat club. Hank Booth was to take over both positions.

At the August 13, 1978 meeting, it was noted that Rev. Donald Duncan would be missed after his tragic and early death. Donations were made to the Steuben Library, Steuben Fire Department, and the Petit Manan Ambulance. The officers elected were Edward "Ted" Denniston, president; Bill Mague, vice-president; Gay Coblens, secretary-treasurer; Peg Reusser, assistant secretary-treasurer; Amy McIlvaine, assistant secretary-treasurer; and Harold Handy, clerk.

The meeting in 1979 was held on August 5, and thanks were extended to Tom and Pete Dardis, Clarkie Wentz, and Dave Brush for the new fence around the cemetery yard. Donations were made to the Steuben Library, Steuben Fire Department, and Petit Manan Ambulance. The officers were re-elected.

From 1980 to 1982, donations were made to the Steuben Library, Steuben Fire Department, and Petit Manan Ambulance. All officers were re-elected.

At the August 14, 1983 meeting, donations were made to the Steuben Library, Steuben Fire Department, and Petit Manan Ambulance. All officers were re-elected, but Priscilla Mague replaced Harold Handy as clerk.

On August 12, 1984, the annual meeting was held. The HAME Corporation donated land next to the cemetery to extend it. Donations were made to the Steuben Library, Steuben Fire Department, and Petit Manan Ambulance. All officers were re-elected.

At the August 18, 1985 meeting, Tom Goettel of the Petit Manan Wildlife Refuge attended to discuss the future plans and issues with the refuge. At this meeting and the one the following year, August 17, 1986, donations were made to the Steuben Library, Steuben Fire Department, and Petit Manan Ambulance. All officers were re-elected.

At the August 9, 1987 meeting, it was announced that church was transferring ownership of the dock to Ted Denniston so it could have proper insurance coverage. It was agreed that money donated in memory of Glenore will be used to purchase twenty new hymnals for the church. Donations were made to the Steuben Library, Steuben Fire Department, Petit Manan Ambulance, and Narraguagus Nursery School. Officers were elected as follows: Ted Denniston, president; Bill Mague, vice-president; Priscilla Mague, clerk; Barbara Snapp, secretary; Frances Mague, treasurer; and Amy McIlvaine, assistant secretary-treasurer.

At the August 14, 1988 meeting, a discussion on the church foundation included a mention that it had badly deteriorated. Work had to be started or the church risked collapse of the tower. A local mason, Danny Barnes, was hired and was in the process of rebuilding the bad sections of foundation. Donations were made to the Steuben Library, Steuben Fire Department, Petit Manan Ambulance, and Narraguagus Nursery School. The slate of officers was re-elected.

At the August 13, 1989 meeting, it was noted that the organ was overhauled and tuned over the winter. Donations were made to the Steuben Library, Steuben Fire Department, Petit Manan Ambulance, and Narraguagus Nursery School. Officers were elected as follows: Ted Denniston, president; Bill Mague, vice-president; Betty Denniston, clerk; Barbara Snapp, secretary; Frances Mague, treasurer; and Amy McIlvaine, assistant secretary-treasurer.

At the 1990–92 meetings, donations were made to the Steuben Library, Steuben Fire Department, Petit Manan Ambulance, and Narraguagus Nursery School. The existing officers were re-elected.

At the August 15, 1993 meeting, the cemetery fence was discussed, and it was reported that the fence was almost complete. The Steuben post office was to take over responsibility for delivery service to Petit Manan once the signs for the 911 emergency street names were in place. Mail delivery had been previously handled by Milbridge. The street names were discussed, and the following were agreed upon: Petit Manan Road for the main road in from Pigeon Hill out to Windswept; Over's Point Road from the triangle between the Magues and section one going out past the church to Over's Point; and Faustina Avenue along the shore from the Armstrong house to Church Street. From Church Street past HAME to the junction with Over's Point was to be named HAME Lane, and Forest Road was the name given to the short piece from Over's Point Road to Faustina Avenue. Donations were made to the Steuben Library, Steuben Fire Department, Petit Manan Ambulance, and Narraguagus Nursery School. The existing officers were re-elected.

In 1994–1996, donations were made to the Steuben Library, Steuben Fire Department, Petit Manan Ambulance, and Narraguagus Nursery School. The existing officers were re-elected.

At the August 10, 1997 meeting, it was noted that the wildlife refuge was going to be placing large rocks on the sides of the gates to prevent people from going around the gates. Carroll West has requested to use three of the moorings and the dock off of Over's Point in exchange for maintaining the road to the dock area. Carroll, one son, and one nephew will be keeping their lobster boats on the moorings. Donations were made to the Steuben Library, Steuben Fire Department, Petit Manan Ambulance, and Narraguagus Nursery School. The officers elected were Ted Denniston and Frances Mague, co-presidents; Amy McIlvaine, vice-president; Michael Denniston, clerk; Barbara Snapp, secretary; and Andrew Booth, treasurer.

At the August 9, 1998 meeting donations were made to the Steuben Library, Steuben Fire Department, Petit Manan Ambulance, and Narraguagus Nursery School. The officers elected were Frances Mague, president; Amy McIlvaine, vice-president; Michael Denniston, clerk; Barbara Snapp, secretary; and Frances Mague, treasurer.

In 1999–2000, donations were made to the Steuben Library, Steuben Fire Department, Petit Manan Ambulance, and Narraguagus Nursery School. The existing officers were re-elected.

At the annual meetings in 2001–02, donations were made to the Steuben Library, Steuben Fire Department, Petit Manan Ambulance, Narraguagus Nursery School, Step-by-Step Day Care, and The Ark. The existing officers were re-elected.

At the annual meeting held August 10, 2003, issues with theft were discussed. This included the wooded donation box bolted to the back shelf, which was presumed to be stolen. It was decided that the organ would be bolted to the floor and that all items of value would be removed after the summer season. Donations were made to the Steuben Library, Steuben Fire Department, Petit Manan Ambulance, Step-by-Step Day Care, and The Ark. A donation was also made to the Over's Cove Corporation, newly formed to support and maintain Over's Point and the dock area. The donation was used to pay off past debt to allow the corporation to start fresh. The existing officers were re-elected.

During the 2004–05 meetings, donations were made to the Steuben Library, Steuben Fire Department, Petit Manan Ambulance, Step-by-Step Day Care, and The Ark. The existing officers were re-elected.

During the 2006–07 meetings, donations were made to the Steuben Library, Steuben Fire Department, Petit Manan Ambulance, Step-by-Step Day Care, and The Ark. The officers were elected as follows: Frances Mague, president and treasurer; Amy McIlvaine, vice-president; and Barbara Snapp, clerk and secretary.

The 2008 meeting was held on August 10, and it was noted that Dave Chapeloney had completed the roofing work. The possibility of continued theft of donations was discussed. Donations were made to the Steuben Library, Steuben Fire Department, Petit Manan Ambulance, Step-by-Step Day Care, The Ark, Cambodian School Project, and the Great Auk Land Trust. The officers were elected as follows: Frances Mague, president and treasurer; Amy McIlvaine, first vice-president; Evans Denniston, second vice-president; and Barbara Snapp, clerk and secretary.

During the August 9, 2009 meeting, donations were made to the Steuben Library, Steuben Fire Department, Petit Manan Ambulance, Step-by-Step Day Care, The Ark, Cambodian School Project, the Great Auk Land Trust, and the Seacoast Mission Food Pantry. The existing officers were re-elected. An official visitor policy as follows was developed:

> Since its founding in 1910, All Souls Church by the Sea has welcomed visitors throughout the year. Services are held each Sunday in July and August, and private services may be arranged for other times by contacting church officials.
>
> Visitors are also welcome to stop in to visit the church at any time, as long as they respect the private property in the area and visit during daylight hours. To protect the privacy of the landowners, visitors are asked to restrict their travels—whether by foot, bicycle or car—to the road leading directly to the church. We encourage visitors to park at the turnaround on the road at the Wildlife Refuge border and walk from there. The church building will be locked during the winter months.

Donations were made to the same slate of organizations at the August 8, 2010 meeting. The following officers were elected: Frances Mague, president and treasurer; Evans Denniston, vice-president; and Barbara Snapp, clerk and secretary.

At the August 7, 2011 meeting, donations were made to the same slate of organizations with the addition of the Downeast Coastal Conservancy. The existing officers were re-elected. A building and maintenance committee was established consisting of Evans Denniston, Jim Kovacs, Bill Mague, and Francie Mague.

From 2012 to 2016, the following organizations were selected for donations at the annual meeting, Steuben Library, Steuben Fire Department, Petit Manan Ambulance, Step-by-Step Day Care, The Ark, Cambodian School Project, the Downeast Coastal Conservancy, and the Seacoast Mission Food Pantry. The existing officers were re-elected.

Over the years, church services were led by either lay people or by clergy, depending on availability. Some of the services were led by Dr. George Foley, Allen R. Van Meter, Mr. Smythe, Rev. Arthur Woolley, Mr. Hall, Mr. Leser, Rev. Donald Duncan, Rev. Mayer, Rev. Omar Patterson, Verdi Reusser, Bob Mitchell, Rev. John Lavoe. Rev. Dana Kennedy, Rev. Woitasek, Jim Kovacs, Bob Simon, Tom McIlvaine, Amy McIlvaine, Barbara Snapp, Frances Mague, Helen and Jeff Panitt, Kathleen Dardis Noe, Jill and Evans Denniston, David Kennedy, Roseanne Dardis, Chris Wehr, Joel and Susie Mague, and Dan Flemming.

The organ has been played over the years by many people, including Gay Coblens, Harriet Heath, Katie Kovacs, Gwen Goode, Betty Denniston, Frances Mague, and Ben Mague.

Older view of All Souls by the Sea.

Original plan of All Souls by the Sea.

A recent view of All Souls by the Sea Church.

6

GETTING THERE IS HALF THE FUN

In the late 1800s and early 1900s, travel was quite a bit different. Today we jump in the car or fly anywhere in the country with relative ease. A century or two ago, getting to Petit Manan Point from places like Philadelphia took a considerable amount of time and money. The easiest travel at that time was by boat, if possible. The second-best choice was by train, and the last choice was by horse and wagon or automobile. Eventually, the roads were improved, making travelling by car the preferred method.

The original plans for the Petit Manan Land Company from July 1889 show two steamship wharfs, one on Sand Point and one at Over's Point. The wharf on Sand Point was never built, but the wharf on Over's Point was and shows up on the tax records from 1896 until 1917. The fact that a steamship wharf was built illustrates that the developers felt the site would be quite large. To support a steamboat wharf, a lot of passengers would need to be using the steamship, or it would not be worth stopping.

No evidence can be found to support that any ships of any size ever used the wharf on a regular basis, and the few people that ended up summering on the point did not justify a stop.

BEFORE 1904

In her journal, Leanora D. Booth described the journey the Denniston family took from Philadelphia in 1902 to get to Petit Manan Point. It started with a train from Philadelphia to New York City, and then taking a different train to Boston, then continuing on to Portland. They took a cab from the Portland train station to the steamboat landing where they boarded SS *Frank Jones*.

SS *Frank Jones* was the sixth ship built by the Bath Iron Works. It measured 253 feet, 2 inches, and drew 7 feet, 6 inches, of draft. It was delivered on August 9, 1892 to the Portland and Machias Company. A side wheel steamship with one smokestack, it travelled from Portland to Machiasport, stopping at points in between, from 1892 to 1904. At that time, it was sold to a shipping line that serviced Long Island Sound route and was moved down there.

After a stop in Bar Harbor and Winter Harbor, SS *Frank Jones* delivered the Denniston family to Wyman just east of Milbridge. The Denniston family was greeted at the dock by Jessie Willie, and they took a buckboard and team over the back roads to Petit Manan Point.

After 1904

After 1904, SS *Frank Jones* stopped running up the Maine coast. A passenger could travel from Philadelphia to Boston on the morning train and catch the evening train out of Boston. The Maine Central had a connection with the Washington County Railroad at Washington Junction, near Ellsworth. The Washington County Railroad was completed at the end of 1898, with the first train running in January 1899 from Washington Junction to Eastport and Calais. The train had stations in Unionville and Cherryfield. A traveler would then take a wagon from Unionville or Cherryfield to Petit Manan Point.

In his book *Home Base*, Ted Denniston described another option available around 1920. That was to take the train out of Philadelphia up through Boston to Bangor and then disembark at Waukeag, the present location of the Route 1 bridge across the narrows in Hancock. From there, the family was greeted by Cora or John Duniver with a Ford Model T. The car and passengers crossed the narrows by ferry. Once across, it took three to four hours to get to Petit Manan Point. About the same time period, Leanora D. Booth wrote that the drive by car was a three-day trip that was shortened to two days as the roads were improved.

A steamboat coming into the wharf at Petit Manan Point.

Schoodic Mountain, Me. On line of W.C.R.R.

The "singing" bridge at Sullivan replaced the ferry and made travel by auto to Petit Manan a little easier.

Opposite above: View of the ferry boat at Sullivan when it was operated by Brad Smith.

Opposite below: A postcard view of the Washington County Railroad with Schoodic Mountain in the background. This was one of the early modes of transportation to get to Petit Manan Point.

7

ALL THE PLAYERS

There were many people over the years who were involved in the development or attempted development of the Petit Manan Point. Others involved who were important influences and owned individual lots are discussed in the Chapters 1 and 4.

SYLVESTER KINSMAN ABBOTT AND REBECCA B. A. ABBOTT

Sylvester K. Abbott, treasurer for the Petit Manan Land Company, was from Malden, Mass. He was born on December 9, 1831 in Bedford. He married Rebecca B. A. Abbott, and they had three sons, John, Ralph, and Harry. Sylvester was listed in the 1880 census as living in Malden, Mass., and working as a book binder.

Death records from 1890 in the City of Malden say Sylvester K. Abbott was a pamphlet binder. His parents are listed as Moses, who was born in Bedford, and Susan, who was born in Billerica.

Rebecca B. A. Abbott died on July 27, 1903. She was the daughter of John Smith and Rebecca Morse and was born in Corinna, Maine. John Smith was born in Washington, NH, and Rebecca Morse was born in Livermore, Maine. Until the time of her death, Sylvester's widow owned lots 102 and 100 in section two of the development. It appears the taxes were not paid after her death and the tax records have a note that the town took the lots in 1911 for 1907 taxes.

FREDERIC R. PAGE

Frederic R. Page was the civil engineer who drew the original plans for the Petit Manan Land Company development in 1889. He was also from Malden, Mass., and this may be why Abbott hired him.

Frederic was the son of Oliver L. and Elizabeth Page. Oliver, who was a store clerk according to the 1870 census, was born in Maine. Frederic's birthplace is listed as

Massachusetts. In the 1870 census, Frederic, who was twenty years old, was a civil engineer and his brother, Joel, who was ten years younger, was in school.

The 1900 census shows Frederic living with his parents. He is listed as head of household. The census also shows that Frederic was unemployed for six months in the previous year. Oliver initially owned what became known as the Dutton Cottage before it was sold to Gardiner H. Dutton.

Gustavus Clark Kilgore and Abbie N. Kilgore

Gustavus C. Kilgore was the treasurer of the Petit Manan Land Company in 1895 and, later, the treasurer of the Petit Manan Land and Industrial Company. He was a doctor in Belfast and owned a pharmacy there.

He formulated and patented the Sarsaparilla compound for the Dana Sarsaparilla Company, located in Belfast. The Sarsaparilla company was bottling 300,000 bottles of the medicine in 1891 when it was sold to a Boston company.

Gustavus C. Kilgore was born on March 9, 1850 to Samuel and Lydia Hinckley Kilgore. Samuel was listed as a wheelwright in the 1850 census for Smithfield, Maine. Gustavus was the next to youngest of Samuel and Lydia's children. The others included Oramandal M., born in 1835; Llewellyn W., born in 1840; Henry L., born in 1841; John M., born in 1844; Florilla Ann, born in 1847; and Selden H., born in 1855.

Gustavus was born in Smithfield, Maine, and he had lived in Belfast for about thirty-seven years at the time of his death on March 29, 1917. He married Abbie N. Ottis on August 6, 1890. In the 1900 and 1910 censuses, he is listed as living with his wife, Abbie O., and nephew, Harry L. Kilgore. He was the city physician of Belfast in 1881–82.

Henry Leonard Kilgore

With the failure of the Petit Manan Land and Industrial Company, Henry L. Kilgore and Frank I. Wilson took ownership of the development. The deed from March 23, 1898 says it includes all the deer contained in the deer park, all map plans, pamphlets, stationary, case of shelves, and other property stored in the stable of G. C. Kilgore in Belfast.

Henry L. Kilgore was born on August 6, 1841 to Samuel and Lydia Kilgore. He died on December 4, 1898, shortly after taking control of the development. This was surely a setback for the Petit Manan Point development. He was the postmaster in the central office in Belfast in 1887 and 1895. He married Helen M. Hussey on October 16, 1879. Helen died on May 22, 1891.

John M. Kilgore

John M. Kilgore was involved in the land development at various stages. He becomes a part of the Maine Coast Club and eventually becomes the sole owner of most of the property.

The death of John M. Kilgore in Northboro, Mass., on Dec. 24, at the age of 87 will be of interest to many of our readers. He was born in Smithfield, Maine, the son of Samuel and Lydia (Hinkley) Kilgore. He was general agent for the Northwestern Insurance Co., in Worcester, Mass., for 30 years and had retired about seven years ago. Mr. Kilgore spent over a quarter of a million dollars during his life on the hotel and development of the Petit Manan Point project. This is, no doubt, a beautiful spot and perhaps if he had been born a half century later when the roads were all rebuilt he might have made a success of his plan or maybe he was a good insurance man and accumulated money fast in that capacity but was not adapted to the hotel or development business. In any event, Mr. Kilgore spent a lot of money here and was a great help to a number of people and if this spot is ever developed to a big summer resort, his pioneer work will ever be remembered. 1931

Obituary of John M. Kilgore from 1931.

John M. Kilgore was born on May 12, 1844 in Smithfield, Maine, to Samuel and Lydia Kilgore. He married Ellen A. Libby on May 6, 1871 in Boston; both were living in Boston at the time. John's occupation was selling life insurance.

Most references refer to his wife as Addie E. or Abbie E. rather than Ellen A. as shown in the marriage record. Addie was the daughter of Daniel and Sarah O. Libby. The 1900 census shows John M., Addie E., and their son, John P. Kilgore, living in Worcester, Mass. Both John M. and John P. were insurance agents. On May 6, 1902, Abbie died in Worcester.

The 1910 census shows John M. living in Worcester with his sons, Fred H. and John P., and daughter, Helen. All three of the men are listed as underwriters for life insurance. The 1920 census again shows John M. Living in Worcester and working as a life insurance agent He died on December 4, 1931, leaving his land to Minnie Casey.

Edgar Filmore Hanson

Edgar F. Hanson was president of the Petit Manan Land Company and, later, the Petit Manan Land and Industrial Company. He was a ten-term mayor of Belfast, a publisher of newspapers and magazines, and the general manager of Dana Sarsaparilla.

He was born on March 11, 1853 in Lincoln, Maine, to Clark and Nancy J. Hatch Hanson. He married Flora E. Nickerson, who died in 1891. He then married Georgie Lord in 1895. He died on April 4, 1933.

In the 1900 census, Edgar is shown as living in Belfast with wife, Georgie; sons, Elvin and Herbert; and daughter, Mabel. His occupation is listed as a real estate agent. In the 1910 census, living arrangements are the same but his occupation is listed as newspaper editor. Georgie's occupation is listed as newspaper proofreader. The 1920 census again shows Edgar living in Belfast with Georgie and working as a newspaper editor. Georgie is listed as a stenographer in the newspaper office.

In 1895 and 1896, Hanson was the president of the Belfast and Moosehead Railroad. He was also one of the first directors of The People's National Bank in Belfast. In January 1897, Hanson launched *The Cream*, a monthly publication containing poetry, stories, and family reading material. Publication was suspended in June 1898 and the Belfast Cream Company went into the hands of a receiver.

Frank I. Wilson

Frank I. Wilson became co-owner of Petit Manan Point property with Henry L. Kilgore on March 23, 1898. He was born on July 17, 1858. It appears he was in the sarsaparilla business with Gustavus Kilgore as well as owner of the pharmacy in Belfast. With Henry L. Kilgore dying shortly after taking control of the property, it appears that Wilson's overall involvement was limited. He transferred his part of ownership to Robert F. Dunton on December 19, 1900.

In the 1920 census for Belfast, he is listed as living in Belfast with his wife, Carrie E.; son, Evan F.; and daughter, Dorris F. He is listed as a collector of insurance. In the 1930 census, he is listed as the tax collector for the city of Belfast. The 1940 census lists him as an insurance agent living with his wife and daughter. He died on February 22, 1943. His wife, Carrie E. Achorn, died on May 14, 1945.

ROBERT F. DUNTON

Robert F. Dunton was the trustee for the Maine Coast Club. He was born on November 24, 1848, the son of Horatio and Julia Ann Dunton in Searsmont, Maine. He was admitted to the bar in 1872 and served as the mayor of Belfast numerous times. He married Elizabeth Mabel Farrar on June 5, 1878. She was born on October 19, 1855.

Dunton is listed in the 1900 census as being a lawyer living with his wife, Lizzie M.; daughters, Helen M., Edith F., and Florence; and son, William F. The 1910 census shows him living with his wife, Elizabeth; daughter, Florence; and son, William. His occupation is again listed as lawyer. The 1920 census lists him as an attorney living with his wife and son.

Elizabeth Dunton died on March 21, 1923, and Robert died on March 3, 1925.

HARRY W. GODDARD

Harry W. Goddard's involvement possibly started as a member of the Maine Coast Club. He owned lots 59–64 in section one. He loaned John M. Kilgore $10,000 in 1923 and then another $5,000 in 1924 as a mortgage against the property.

Goddard was born on September 14, 1863 in Holyoke, Mass., to Dorrance S. and Mary H. Goddard. He married Georgia Grace Watson on December 14, 1887. In the 1900 census, he is listed as living in Spencer, Mass., with his wife, Georgia, and daughters, Eleanor G. and Marion W. He was working in the wire manufacturing business. In the 1910 census, he is listed as living in Worcester, Mass., with his wife and daughters and working as a wire company employer. The 1920 census has him living with his wife and working as a manufacturer of wire. In the 1930 census, G. Grace Goddard is shown as widowed, living in Worcester.

ELEANOR GODDARD DANIELS

With Harry W. Goddard's death before the repayment of the debt by John Kilgore, Eleanor G. Daniels became involved with Petit Manan Point.

Eleanor was born on March 25, 1889 in Spencer, Mass., to Harry W. and Georgia Grace Goddard. On June 2, 1915, she married Fred Harold Daniels, who was listed as a sales engineer on the marriage license. The couple appears in the 1920 census as living in Worcester, Mass., with their daughter, Eleanor. His occupation is listed as general manager of Riley Stoker. Eleanor would go on after her father's death to own most of Petit Manan Point and operate Petit Manan in Maine with the Halls before selling the property to William Mague.

Ferdinand Hess Shumaker

Ferdinand H. Shumaker became involved in Petit Manan Point when he purchased the property from John M. Kilgore in 1926.

Shumaker was born in Ortonville, Minn., on June 23, 1890. He was the son of Abraham Lincoln Shumaker and Kate Hess Shumaker. His World War I draft registration card from May 31, 1917 shows him living in South St. Paul, Minn.

He married Helen Marion Sorensen in Manhattan, NY, on April 21, 1923. She was the daughter of Neil Sorensen and Mary Smith Sorensen.

The mortgage for the property, which is also signed by his wife, states that he is from North Kennebunkport, York County, Maine.

In the 1940 census, Ferdinand H. and Helen M. are shown living in Teaneck Township, NJ, and he was working as the solicitor for a trucking company. Ferdinand's World War II draft registration card shows him still living in Teaneck, and working for Fishers Express in Lodi, NJ. Ferdinand died in Texas on September 11, 1985.

Winthrop Goddard Hall and Madeline Tabor Hall

Winthrop G. Hall and Madeline T. Hall operated Petit Manan in Maine Inc. with Winthrop's cousin, Eleanor G. Daniels. Winthrop was the son of Charles S. Hall and Emma Goddard. Madeline was the daughter of Edward P. Tabor and Jane B. Dickinson.

The 1900 census shows Charles and Emma living in Worcester, Mass., with sons, Winthrop G. and Roderic C., and daughters, Annie F., Margaret W., and Elizabeth B. Charles' occupation is listed as bookkeeper.

Winthrop and Madeline were married on June 19, 1906. The 1910 census shows Winthrop G. Hall and Madeline T. Hall living in Worcester with son, Stephen, and Winthrop's occupation is listed as assistant superintendent of a motor company.

On Winthrop's World War I draft registration card from 1918, his occupation is listed as superintendent at Spencer Wire Company.

The 1930 census shows Winthrop and Madeline with living in Worcester with son, Stephen G., and daughters, Margaret and Priscilla. Winthrop's occupation is listed as sales manager at an automobile dealer. The 1940 census shows Winthrop and Madeline still in Worcester with his occupation listed as salesman for a motor company. Madeline died in 1968 and Winthrop died in 1977.

William Robert Mague and Priscilla Hall Mague

William R. Mague was born in 1908 and married Priscilla W. Hall, the daughter of Winthrop and Madeline Hall. He was the son of William E. and Florence Mague, who are listed on the 1910 census as living in Plainfeield, NJ, with their son and daughter. William's occupation is listed as machinist. In the 1920 census, William E. and Florence A. Mague are in Worcester with their son, William R., and daughter, Florence G. William E. Mague's occupation is listed as a machinist at a wire mill. The 1930 census lists the couple with son, William R., in Worcester. William's occupation as a machinist at a steel mill.

Both William and Priscilla attended Westminster Choir College in Princeton, NJ, in the early 1930s. Westminster Choir College was started in 1926 to educate students at the undergraduate and graduate level for careers in music education and other music-related careers.

The Magues became ministers of music at New Haven and Springfield, and they ran a music camp at Petit Manan for members of the choirs of those cities. The students who attended the camps stayed in the clubhouse building and the music classes were held in the All Souls by the Sea Church.

The Mague family moved to Petit Manan Point full time in 1946. The Magues worked at All Souls Congregational Church in Bangor, Maine, and William became a teacher at the Northern Conservatory of Music, founded in 1929 at 166 Union St., in Bangor in the Isaac Farrar Mansion. William R. Mague served as director of the Northern Conservancy of Music from 1970 until 1972, when it closed.

William R. Mague died in 2002 and Priscilla died on October 21, 1987.

8

A National Wildlife Refuge

The following is a brief history of Petit Manan Point National Wildlife Refuge. This deals only with the land on Petit Manan Point and not all the additional land and islands that the wildlife refuge operates. A lot of the information was taken from the annual narrative reports that the wildlife refuge produced from 1983 to 1997.

The official change of ownership took place on October 14, 1975 as a partial donation by Mr. and Mrs. William Mague.

A news release by the Department of the Interior, Fish and Wildlife Service on October 12, 1975 announced the acquisition of the 1,700-acre peninsula, saying it was good habitat for eider ducks and black ducks, as well as other migrating waterfowl. The release stated that the property would become part of the National Wildlife Refuge System and officially designated it the Petit Manan National Wildlife Refuge.

A *New York Times* article from October 14, 1975 calls the property "one of the largest undeveloped peninsulas on the Maine coast."

The official transfer ceremony was held the evening of October 14, 1975 at the H. D. Moore Parish House in Steuben.

In 1978, Douglas Mullin was the manager of the refuge.

In 1983, the Petit Manan National Wildlife Refuge was being managed out of the Moosehorn National Wildlife Refuge. Petit Manan did not have a budget or any staff at this point. Public use was not encouraged but a small parking lot, bulletin board, and a short trail to the shore were built.

On December 19, 1984, Assistant Refuge Manager Thomas A. Goettel transferred from Moosehorn to the Milbridge area to work full time on the island satellites. He worked out of his house initially. It was noted that the whitetail deer and coyotes were flourishing on the peninsula, probably due to an increase in law enforcement in 1984. There were two small trails at the parking lot near the refuge entrance at this time, and in 1984, both were brushed and blazed, and all signs were repainted.

The first year that Petit Manan National Wildlife Refuge had its own operating budget was 1985. The loop trail around Birch Point was added that year. The staff consisted of Goettel and two students, Mike Zeiser and James L. Burney, who worked

Sign on U.S. Route 1 directing people to Petit Manan Point.

for two months as student conservation association volunteers. Burney then served a biological aid from September to November, when he left. A refuge office was built in Milbridge by a local developer and rented. The one-room building was 20 feet by 30 feet and had a garage. In September, the refuge moved in—and out of Goettel's residence.

On May 28, 1986, refuge representatives were talking to the Magues about their remaining 175-acre tract, which the Magues were considering donating. That year, the staff included Goettel, plus temporary Biological Technician Kevin Reyor and volunteers Daniel Ottman and Ralph Widrig. The trails to Birch Point and Big Pond were improved during the year.

In 1987, Goettel was refuge manager. Other staff included Daniel Ottman, who was a summer biological aid, and Ralph S. Widrig, a part-time biological technician. It was noted that Widrig was paid for two days a week and volunteered the other days.

A number of the fields at the Point were burned for fire management, including the Birch Clearing-South Field, parking lot field, and the Meadow Brook Field. The annual report noted that many of the fields had not been burned in ten to twenty years. It was also noted that Priscilla Mague died suddenly on October 21 at her home on the Point and that her love for and support of the refuge would always be missed.

In 1988, the Skinners, who owned Windswept, entered into talks with the refuge about selling their property. The Skinners wanted to see the property become part of the refuge, but they could not agree on a price and sold the property to their daughter. The staff that year included Goettel, David Schick as maintenance mechanic, and William Vasquez and Ralph S. Widrig as biological technicians. It is also noted that Vasquez and his three children, Cathy, Matt, and Dina, volunteered many hours at the refuge.

The existing dams built for controlling water for cranberry production and beaver dams were surveyed. Replacement dams were designed for Meadow Brook and Mague, and the refuge applied for proper permits. The Silver Mine Field, Birch Clearing-North, and the Camp Roads field were burned for fire management. A new footbridge over a small stream on the Shore Trail was built.

In 1989, the staff consisted of Goettel; David Schick, maintenance mechanic; William Vasquez and Ralph Widrig, biological technicians; Russell Worcester and Elizabeth Demetrius, motor vehicle operators; and Richard Schauffler, Glen Mittelhauser, and George Baggett, maintenance workers. Most of the positions were temporary and only lasted a few months. Volunteers included Don and Karis Herriott and Karen Koch. Work was started on new dams for the Mague and Meadow Brook areas, but snow on November 21 stopped the work for the year. In July, new footbridges were built over three wet spots on the Birch Trail.

The 1990 staffing included Goettel; Alvah E. Fickett, Jr., and Russell Worchester, motor vehicle operators; and Richard Schauffler, David Schick, and David Hiltz, maintenance workers. New trail head signs were installed for both the Birch Trail and the Shore Trail. Both of the dams at Mague and Meadow Brook were completed after being started in the previous year.

After fifteen years without any signs, the refuge received funding for entrance and interpretive signs, which were installed that year. John and Karen Hollingsworth visited the refuge in August to take photos for their photographic essay of the National Wildlife Refuge System.

In 1991, the refuge changed leadership. Goettel was transferred to Eastern Neck National Wildlife Refuge and Margaret M. Anderson became refuge manager in June. Maurice Mills served as acting refuge manager in the interim. Other staff included David Schick, maintenance mechanic; Jodie L. Kimmel, office assistant; Ralph Widrig and Norm Famous, biological technicians; Karyn Kocincki, student trainee; and Ritchie Hafford and Myron Alley, motor vehicle operators.

The year 1991 also marked the first time Petit Manan was independent rather than a satellite of Moosehorn National Wildlife Refuge.

A photo in the annual report of refuge personnel included Bill Mague, with the caption reading:

> Bill Mague is the reason we have Petit Manan National Wildlife Refuge. His vision and incredible persistence ensured protection of this unique peninsula. A longtime resident, he owned what is now the mainland portion of the refuge. Over the years he has been unfailing in his support of refuge projects. Bill utilizes his own equipment to mow hayfields, roadsides, and to maintain and plow the road, quite remarkable for a man in his eighties. Perhaps more telling is the way he has met the sometimes "bizarre" antics of how we go about conducting business in the government with remarkable forthrightness and truly gentlemanly demeanor. He provides historical perspective and suggestions or solutions to current projects. We could not run the refuge without him.

In 1991, Paul Wehr placed a development easement on his 15 acres bordering the refuge.

In 1992, staff included Anderson; Brian E. Benedict, supervisory refuge operations specialist; David Schick, maintenance mechanic; Jodie L. Kimmel, office assistant; Jennifer Megyesi, co-op biologist trainee; and Norn Famous and Ralph Widrig, biological technicians. It was noted in the annual report that Bill Mague allowed the refuge to store boats and equipment in his barn over the winter. Some volunteers completed a service project that involved nailing the roof on the Mague barn and preparing the old barn for burning. The blueberry fields were part of a prescribed burn.

The Cranberry Dam was completed in 1993. The Birch Point Parking Lot Field, Meadow Brook Field, Shore Tail Field, South Birch Point Field, and the Lower Hayfield were part of a prescribed burn. Staff included Anderson; Benedict; David Schick, small craft operator; Jodie L. Kimmel, office assistant; and Jennifer Megyesi, co-op biologist trainee.

The 1994 staff was unchanged from the previous year. The Powerline Field and the Homestead Field were part of prescribed burns. The Birch Point Parking area was evaluated for paving and to begin work on engineering drawings.

In 1995, a new refuge manager, Stanley A. Skutek, took over in July when Anderson was transferred to Agassiz National Wildlife Refuge in Minnesota. Benedict remained as supervisory refuge operations specialist with the rest of the staff, including David Schick, small craft operator; Jodie L. Kimmel, office assistant; Jennifer Megyesi, refuge biologist; and Eric Morrison, biological technician.

The boardwalk on the Shore Trail was built and discussions began on the possible renaming of the Shore Trail in honor of John Hollingsworth, Jr., who died in June 1995. A half-mile loop addition to the Shore Trail was roughed out by volunteers. At the information kiosk, a handicapped accessible parking area was installed, consisting of a gravel base to be topped with a stone dust the following year.

The 1996 staff included Skutek; Benedict; Schick; Kimmel; Stephanie Koch, refuge biologist; and Eric Morrison, contract laborer. The Shore Trail was renamed the John Hollingsworth Memorial Trail. Hollingsworth and his wife, Karen, travelled to more than 400 refuges and photographed them. At the end of 1995, the refuge was honored to learn that it had been chosen as John and Karen Hollingsworth's favorite refuge.

The final layer of stone dust was added to the handicap parking area and compacted. Topping gravel was added to Petit Manan Road and large boulders were added to the sides of the gates to prevent people from driving around them when they were closed. The refuge also acquired Abbott Island, located in Carrying Place Cove, which was most likely named for Sylvester Abbott.

The 1997 staff remained unchanged from the previous year. On October 18, 1997, the John Hollingsworth Memorial Trail was officially dedicated. The refuge staff moved into a new office in Milbridge, where it is currently located.

In February 2004, Charlie Blair took over as refuge manager when Stanley Skutek moved to Canaan Valley National Wildlife Refuge. In December, the name Petit Manan National Wildlife Refuge was officially changed to Maine Coastal Islands National Wildlife Refuge to better express the mission and scope of the refuge that had grown to include forty-six islands off the coast of Maine.

In 2006, the refuge opened 655 acres at Petit Manan to muzzleloader whitetail deer hunting from November 27 to December 2. A total of seventy-three participated in the hunt. This was the first-time hunting had been permitted since before 1976, when the refuge had been established. No deer were harvested during the 2006 season.

In 2009, the refuge staff installed 200 feet of boardwalk, 880 feet of drainage and six water bars, and rerouted 700 feet of the Birch Point Trail. Work was done by refuge staff Jim Fortier and Craig Smith along with student interns. In 2009, Beth Goettel was the refuge manager.

In 2010, interns through the Student Conservation Association completed the work that was started in 2009 on the Birch Point Trail. An additional 444 feet of boardwalks was built, as well as 2,000 feet of raised surface. In addition, 150 feet of drainage pipe were installed, and 150 feet of trails were cleared of brush and benches were placed along the trail. Beth Goettel was still the refuge manager.

In 2014, the refuge had grown to protect sixty islands. A cleanup effort was initiated at Petit Manan Point by refuge staff and volunteers, who cleared and 60 cubic yards of debris weighing 8,680 pounds from the shore at the point.

In 2019, as of this writing, Brian Benedict is the refuge manager; Eddy Edwards, deputy refuge manager; Linda Welch, supervisory wildlife biologist; Michael Langlois and Sara Williams, wildlife biologists; Teressa Cultrera, administrative assistant; James Fortier, boat operations and maintenance; Roger Chandler, maintenance; and Sean Billings, volunteer trail tender.

A view of Cable Cove named because the telephone cable from the mainland out to Petit Manan Lighthouse went into the water here. The cable can be seen to the right of the photo.

The cable laying at Cable Cove from when it ran out to communicate with Petit Manan Lighthouse.

Adirondack chairs located at Lobster Point provide a nice rest after the hike.

Bear Cove on the western side of the refuge is one of the harder places to reach but being there is almost a holy experience.

A spruce grouse along the Hollingsworth Trail at Petit Manan Point.

Opposite above: A snowshoe hare at Petit Manan Point.

Opposite below: Porcupines are very plentiful at Petit Manan Point.

A crab in a tidal pool at Petit Manan Point.

A deer walking through a blueberry field that was part of the original deer park.

We hope you enjoyed this book, but please respect private property and do not disturb any of the historic areas.

Bibliography

Tax Records, Town of Steuben, 1865, 1871–1882, 1884–1886, 1888–1890, 1893–1894, 1896–1923, 1925–1934, 1936–1961

All Souls by the Sea Annual Meeting Minutes, 1957 to 2016

Denniston, N., Journal of Nancy Denniston—Petit Manan Point, July 31, 1965

Denniston, T., Journal of Ted Denniston, *Home Base*, 1993

Booth, L., Journal of Leanora D. Booth, *Petit Manan, 1900–1968*

Morrison, H., *Ownership History of "Whileaway" Cottage, Petit Manan Point, Maine*, Homer Morrison, October 05, 2009

Petit Manan National Wildlife Refuge, Annual Narrative Reports, 1985 to 1997

Maine Coastal Islands National Wildlife Refuge, Refuge Highlights, for years 2004, 2005, 2006, 2009, 2010, 2014

Coblens, G., *All Souls Church by the Sea, Petit Manan Point, Maine*, Glenore Orris Coblens, 1972

Newspaper Articles

"Rival of Bar Harbor, Rapid Development of Petit Manan," *Boston Daily Globe*, pp. 2, June 16, 1889

Bangor Daily Whig and Courier, June 15, 1889

"Immense Deer Park, to be one of the features of a Maine Resort," *Bangor Daily Whig and Courier*, December 9, 1889

"Washington County Column," *Bangor Daily Whig and Courier*, January 29, 1890

"The Deer Park," *Bangor Daily Whig and Courier*, February 13, 1890

"Deer Park," *Bangor Daily Whig and Courier*, May 13, 1890

"Capt. F. O. Parker at Petit Manan," *Bangor Daily Whig and Courier*, May 21, 1895

"Mayor E. F. Hanson and Dr. G. C. Kilgore," *Bangor Daily Whig and Courier*, July 13, 1895

"Steamer Elm City," *Bangor Daily Whig and Courier*, September 16, 1895

"Petit Manan Land and Industrial Company," *Bangor Daily Whig and Courier*, October 8, 1895

"Dr. G. C. Kilgore and Mayor E. F. Hanson," *Bangor Daily Whig and Courier*, Octo 2, 1895

"James T. Pottle of Belfast has drawn plans," *Bangor Daily Whig and Courier*, November 11, 1895

"Work of every description is being literally 'rushed' at the new resort," *Bangor Daily Whig and Courier*, April 20, 1896

"Hearing at Augusta," *Bangor Daily Whig and Courier*, October 27, 1896

"A peculiar case of search and seizure," *Bangor Daily Whig and Courier*, Octo 30, 1896

"A hearing was held," *Bangor Daily Whig and Courier*, February 19, 1897

Bangor Daily Whig and Courier, Petit Manan Case, June 8, 1897

"R. W. Dunn Fails," *Boston Daily Advertiser*, Issue 36, col. C, February 11, 1898

"Petit Manan Land Company Sold out," *Boston Daily Advertiser*, Issue 62, col. B, p. 2, March 14, 1898

"Bar Harbor's Merry Season," *New York Times*, p. 16, July 7, 1901

"The Maine Coast Club," *New York Times*, p. 1, March 17, 1902

"Sportsman's Paradise," *Daily Kennebec Journal*, April 4, 1902

"Stock and Real Estate," *American Angler Advertiser*, p. 6, June 1896, Wm. C. Harris Editor, New York City

"8 Per Cent Investment, Principal secured under deed of trust to International Trust Co.," *Recreation*, p. 22

Brochures

Petit Manan, the New Resort on the Coast of Maine, Petit Manan Land Company, sixteen pages, *circa* 1888

Kilgore, John M. Kilgore, 4 pages

Penzance in Maine, Ferdinand H. Shumaker, trifold, *circa* 1927

Petit Manan in Maine, Mr. and Mrs. Winthrop Hall, six-page trifold

Petit Manan in Maine, Mr. and Mrs. Winthrop Hall, four-page bifold

Land Plans

Plan of Township 4 by Lothrop Lewis Esq. January 8, 1794 and updated August 28, 1869, on file at Washington County Courthouse

Map of Properties of the Petit Manan Land Company, Steuben, Washington County, Maine, Frederic R. Page, Malden, Mass.

Plan of Villa Lots, Petit Manan Land Company near Bar Harbor Maine, Frederic R. Page, July 1889, Revised February 1895, Section 1, on file at Washington County Courthouse, Vol. 1, Page 13

Plan of Property of Maine Coast Club, Petit Manan Point, Steuben, Maine, Revised 1907, Section 1

Plan of Villa Lots, Petit Manan Land Company near Bar Harbor Maine, Frederic R. Page, February 1895, Section 2, on file at Washington County Courthouse, Vol. 1, Page 17

Plan of Villa Lots, Petit Manan Land Company near Bar Harbor Maine, Frederic R. Page, July 1889, Revised February 1895, Section 3, on file at Washington County Courthouse, Vol. 1, Page 15

Plan of Villa Lots, Petit Manan Land Company near Bar Harbor Maine, Frederic R. Page, April 1895, Addition to Sections 2 and 3, on file at Washington County Courthouse, Vol. 1, Page 16

Plan of Villa Lots, Petit Manan Land Company near Bar Harbor Maine, Frederic R. Page, February 1895, Section 4, on file at Washington County Courthouse, Vol. 1, Page 14

Plan of Yellow Birch Head Property, Petit Manan Point, Steuben, Maine, Daniel to Stauffer, May 1943 by W. G. Hall, on file as Washington County Courthouse

Plan of Petit Manan Industries, Inc., Subdivision located at Hilton Head off Petit Manan Road, C. A. Mitchell, 1972, on file at Washington County Courthouse, Book 7, Page 1A

Deeds and Mortgages

Deed, Township 4 to Thomas Ruston, August 26, 1794, Vol. 1, Page 391

Deed, Thomas Ruston to Samuel Freeman, April 22, 1795, Vol. 1, Page 489

Foreclosure, Samuel Freeman, May 25, 1805, Vol. 4, Page 89

Deed, Samuel Freeman to William Freeman, November 2, 1826, Vol. 39, Page 190

Deed, Thomas Palmer to Asa, Freeman, Vinas Torry, April 27, 1859

Deed, Asa, Freeman, Vinas Torry to Thomas Palmer, April 20, 1859, Vol. 93, Page 446

Deed, Nathan Hinkley to Asa and Vinas Torry, May 4, 1863, Vol. 103, Page 282

Deed, Asa and Vinas Torry to Nathan Hinkley & Albion Hinkley, November 5, 1863, Vol. 104, Page 565

Deed, Nathan Hinkley & Albion Hinkley to James Mansfield, January 22, 1864, Vol. 108, Page 280

Deed, James Mansfield to Nathan Hinkley, January 22, 1864, Vol. 144, Page 193

Deed, Warren Leighton to James Mansfield and Nathan Hinkley, November 18, 1864, Vol. 138 Page 5

Deed, Nathan Hinkley to Andrew James, Everand Greeley, David W. Campbell, William C. Ricker, September 10, 1874, Vol. 138, Page 60

Deed, James Wallace, William Godfrey, Franklin Brown to Everard H. Greely, William C. Ricker, Andrew S. James, September 10, 1874, Vol. 138, Page 62

Deed, Mary Smith to Sylvester K. Abbott, Vol. 173, Page 334

Deed, Charles J. Smith to Sylvester K. Abbott, Vol. 173, Page 333

Deed, William C. Ricker, Emily D. Ricker, Andrew S. James, Ina M. James, Charles Campbell, Frank Campbell, Ella Campbell, Abby W. Campbell, M.W. Patten, F.W. Patten, Everard H. Greely, Ann F. Greely, John W. Hale Jr., Rebecca W. Hale, George W. Hunt, Eliza Hunt to Sylvester K. Abbott, June 3, 1881, Vol. 157, Page 60

Deed, Ezra Over, Nancy S. Over, Abbie Over to Sylvester K. Abbott, August 12, 1881, Vol. 157, Page 184

Deed, Sylvester K. Abbott to Petit Manan Land Company, April 19, 1889, Vol. 186, Page 113

Deed, Petit Manan Land Company to Petit Manan Land & Industry Company, November 21, 1895, Vol. 213, Page 385

Indenture, Petit Manan Land & Industry Company to International Trust Company, December 26, 1895, Vol. 213, Page 396

Indenture, Petit Manan Land & Industry Company to International Trust Company, Vol. 213, Page 402

Deed, Petit Manan Land & Industry Company to John H. Flannagan, November 9, 1897, Vol. 212, Page 78

Deed, John H. Flannagan to William O. Alden, February 21, 1898, Vol. 225, Page 282

Deed, Petit Manan Land & Industry Company to Henry L. Kilgore and Frank I. Wilson, March 23, 1898, Vol. 224, Page 448

Deed, William O. Alden to Charles A. Wilber, December 16, 1898, Vol. 232, Page 26
Deed, Estate of Henry Kilgore to Robert F. Dunton, December 19, 1900, Vol. 230, Page 511
Deed, Frank I. Wilson to Robert F. Dunton, December 19, 1900, Vol. 230, Page 513
Deed, Charles A. Wilber to Robert F. Dunton, December 19, 1900, Vol. 240, Page 232
Deed, Everand H. Greely to Robert F. Dunton, December 19, 1900, Vol. 230, Page 508
Deed, Maine Coast Club, Robert F. Dunton Trustee to 1st National Bank of Madison, December 31, 1906, Vol. 283, Page 222
Deed, Robert F. Dunton to Henry H. Gray and Charles H. Clark in trust, January 21, 1907, Vol. 280, Page 5
Deed, Augusta Trust Company to John M. Kilgore, June 10, 1907, Vol. 273, Page 507
Deed, KC Gray to John M. Kilgore, June 10, 1907, Vol. 273, Page 506
Request to transfer, Charles H. Clark to Frank W. Sawyer trustee for John M. Kilgore, July 10, 1907, Vol. 283, Page 223
Deed, Charles H. Clark to Frank W. Sawyer trustee for John M. Kilgore, July 23, 1907, Vol. 280, Page 117
Deed, Abbie O. Kilgore, Georgia G. Hanson, Gustavus C. Kilgore, Edgar F. Hanson to John M. Kilgore, November 19, 1908, Vol. 275, Page 145
Deed, Frank W. Sawyer to John M. Kilgore, December 10, 1908, Vol. 283, Page 224
Deed, John M. Kilgore to Lester Bailey, Eleanor Denniston, John M. Kilgore, January 6, 1914, Vol. 298, Page 456
Indenture, John M. Kilgore to A.L. Stewart & Sons, January 4, 1922, Vol. 348, Page 517
Mortgage, John M. Kilgore to Charles A. and Frank H. Stewart, January 10, 1922, Vol. 354, Page 77
Mortgage, John M. Kilgore to Harry W. Goddard, May 25, 1923, Vol. 359, Page 68
Mortgage, John M. Kilgore to Harry W. Goddard, February 8, 1924, Vol. 359, Page 228
Mortgage, Ferdinand H. Shumaker to John M. Kilgore, May 29, 1926, Vol. 372, Page 311
Notice of Foreclosure, John M. Kilgore to Ferdinand Shumaker, Vol. I, Page 457
Assignment, Kilgore from Shumaker, May 4, 1929, Vol. 379, Page 463
Notice of Foreclosure, Charles A. and Frank H. Steward to John M. Kilgore, February 23, 1928, Vol. I, Page 410
Mortgage Sale, Charles A. and Frank H. Steward to Mechanics National Bank of Worcester, January 10, 1929, Vol. 379, Page 348
Mortgage, John M. Kilgore to Minnie C. Casey, January 19, 1929, Vol. 379, Page 354
Deed, Mechanics National Bank of Worcester to Petit Manan in Maine Inc., January 25, 1930, Vol. 377, Page 582
Deed, Mechanics National Bank of Worcester to Petit Manan in Maine Inc., January 25, 1930, Vol. 377 Page 583
Deed, Mechanics National Bank of Worcester to Petit Manan in Maine Inc., December 15, 1930, Vol.387, Page 191
Deed, Mechanics National Bank of Worcester to Eleanor G. Daniels, April 16, 1940, Vol. 441, Page 6
Deed, John M. Kilgore to G. Horton Glover, Vol. 250, Page 8
Will, John M. Kilgore, 1932, Vol. W-4, Page 162
Deed, Leonard R. Hillgrove to William Sawyer, January 3, 1957, Page 548, Vol. 23
Deed, Austin Shaw to William Sawyer, April 29, 1976, Vol. 923, Page 100
Deed, Eleanor G. Daniels to William R. Mague, March 6, 1950, Vol. 494, Page 335
Deed, Eleanor G. Daniels to Petit Manan in Maine, Inc., March 7, 1950, Vol. 494, Page 334

Deed, William R. Mague to Petit Manan Industries, Inc., October 14, 1975, Vol. 901, Page 243
Deed, Petit Manan Industries, Inc. to Nature Conservancy of the Pine Tree State, October 8, 1975, Vol. 901, Page 248
Deed, Petit Manan Industries, Inc. to Nature Conservancy of the Pine Tree State, October 8, 1975, Vol. 905, Page 46
Deed, William Sawyer to Nature Conservancy of the Pine Tree State, July 13, 1976, Vol. 931, Page 221
Deed, William Sawyer to Nature Conservancy of the Pine Tree State, July 13, 1976, Vol. 931, Page 224
Deed, Nature Conservancy of the Pine Tree State to United States of America, June 28, 1976, Vol. 931, Page 232
Deed, Nature Conservancy of the Pine Tree State to United States of America, August 19, 1976, Vol. 937, Page 35
Deed, Miranda P. Stanhope to Freeman Leighton, John McNamara, Warren Leighton, August 28, 1857, Vol. 116, Page 278
Deed, John McNamara to Warren Leighton, February 19, 1863, Vol. 116, Page 279
Deed, Warren Leighton to James Mansfield and Nathan Hinkley, November 18, 1864, Vol. 138, Page 5
Deed, Frances S. Sawyer to Leonard R. Hillgrove, May 29, 1951, Vol. 505, Page 96
Deed, Leonard R. Hillgrove to William Sawyer, January 3, 1957, Vol. 548, Page 23
Deed, Austin Shaw to William Sawyer, April 29, 1976, Vol. 923, Page 100
Deed, Township 4 to Samuel Kingsley, June 25, 1795, Vol. 1, Page 480
Deed, Samuel Kingsley to Andrew Dyer, June 1, 1799
Deed, Andrew Dyer to Reuben Dyer, April 12, 1806, Vol. 5, Page 344
Deed, Andrew Dyer to Reuben Dyer, April 1, 1808, Vol. 5, Page 345
Deed, Reuben Dyer to Wilson M. Dyer, May 29, 1845, Vol. 56, Page 66
Deed, Wilson M. Dyer to Henry Over, August 6, 1850, Vol. 70, Page 8
Deed, Henry Over to Nancy Over, July 7, 1874, Vol. 136, Page 439
Deed, Ezra Over to Nancy Over, July 7, 1874, Vol. 136, Page 438
Deed, Nancy Over to Ezra Over, July 7, 1874, Vol. 136, Page 438
Deed, Ezra Over, Nancy S. Over, Abbie Over to Sylvester K. Abbott, August 12, 1881, Vol. 157, Page 184
Deed, Petit Manan Land Company to Badger & Manley, Vol. 213, Page 220, Lot 87, Section 1
Deed, Petit Manan Land Company to E. N. Carver, Vol. 213, Page 232, Lot 117, Section 4
Deed, Petit Manan Land Company to C. Thornton Libby, Vol. 208, Page 365, Lot 55, Section 1
Deed, Petit Manan Land Company to Willis L. Watson, Vol. 208, Page 413, Lot 245, Section 4
Deed, Petit Manan Land Company to Rebecca B. A. Abbott, Vol. 218, Page 288, Lot 102, Section 2
Deed, Petit Manan Land Company to L. L. Bateman, Vol. 213, Page 291, Lot 199, Section 4
Deed, Petit Manan Land Company to Charles B. Bradford, Vol. 213, Page 381, Lot 125, Section 3
Deed, Petit Manan Land Company to Burleigh Flynt, Vol. 218, Page 216, Lot 83, Section 1
Deed, Petit Manan Land Company to Elmira A. Champlain, Vol. 213, Page 375, Lot 101, Section 2
Deed, Petit Manan Land Company to Linus Darling, Vol. 213, Page 288, Lot 113, Section 2
Deed, Petit Manan Land Company to H. L. Goodwin, Vol. 230, Page 2, Lot 91, Section 2
Deed, Petit Manan Land Company to Mary L. Gray, Vol. 213, Page 280, Lot 37, Section 2
Deed, Petit Manan Land Company to Horance N. Hastings, Vol. 224, Page 35, Lot 42, Section 1
Deed, Petit Manan Land Company to Horance N. Hastings, Vol. 224, Page 36, Lot 85, Section 1
Deed, Petit Manan Land Company to W. J. Heffernan, Vol. 213, Page 286, Lot 173, Section 3

Deed, Petit Manan Land Company to Mac Kendall, Vol. 213, Page 285, Lot 211, Section 2
Deed, Petit Manan Land Company to A. L. Ladd, Vol. 224, Page 23, Lot 157, Section 4
Deed, Petit Manan Land Company to W. M. Ladd, Vol. 224, Page 25, Lot 158, Section 4
Deed, Petit Manan Land Company to J. H. Livingston, Vol. 224, Page 76, Lot 49, Section 1
Deed, Petit Manan Land Company to A. J. Lumley, Vol. 224, Page 138, Lot 185, Section 4
Deed, Petit Manan Land Company to Maine Outing Club, Vol. 224, Page 70, Lot 261, Section 4
Deed, Petit Manan Land Company to H. Mason & Son, Vol. 213, Page 258, Lot 103, Section 2
Deed, Petit Manan Land Company to J. J. MacDonald, Vol. 224, Page 400, Lot 47, Section 1
Deed, Petit Manan Land Company to J. J. MacDonald, Vol. 224, Page 401, Lot 46, Section 1
Deed, Petit Manan Land Company to Newtown Journal, Vol. 213, Page 272, Lot 77, Section 1
Deed, Petit Manan Land Company to Opinion Publishing Company, Vol. 224, Page 55, Lot 99, Section 1
Deed, Petit Manan Land Company to Oliver L. Page, Vol. 226, Page 337
Deed, Petit Manan Land Company to J. J. Parker, Vol. 213, Page 338, Lot 17, Section 1
Deed, Petit Manan Land Company to Petit Manan Land & Industrial Company, Vol. 213, Page 385
Deed, Petit Manan Land Company to J. D. Sale, Vol. 213, Page 251, Lot 133, Section 1
Deed, Petit Manan Land Company to J. D. Sale, Vol. 213, Page 252, Lot 311, Section 3
Deed, Petit Manan Land Company to P. L. Schriftgresser, Vol. 224, Page 21, Lot 147, Section 3
Deed, Petit Manan Land Company to H. A. Shorey & Son, Vol. 230, Page 23, Lot 142, Section 2
Deed, Petit Manan Land Company to F. J. Smith, Vol. 224, Page 290, Lot 569, Section 3
Deed, Petit Manan Land Company to Sun Publishing Company, Vol. 229, Page 3
Deed, Petit Manan Land Company to B. B. Thatcher, Vol. 218, Page 313, Lot 10, Section 1
Deed, Petit Manan Land Company to Charles E. Waterman, Vol. 213, Page 249, Lot 137, Section 4
Deed, Petit Manan Land Company to Will Watson, Vol. 213, Page 256, Lot 247, Section 4
Deed, Petit Manan Land Company to Hattie F. Whipple, Vol. 213, Page 488, Lot 75, Section 2
Deed, Petit Manan Land Company to F. J. Wilson, Vol. 213, Page 358, Lot 23, Section 1
Deed, Petit Manan Land Company to George L. Wing, Vol. 213, Page 277
Deed, Petit Manan Land Company to Wm T. Howard, Vol. 248, Page 45, Lot 115, Section 3
Deed, Petit Manan Land Company to W. E. Redding, Vol. 230, Page 476, Lot 289, Section 4
Deed, Petit Manan Land Company to The Transcript Company, Vol. 230, Page 534, Lot 53, Section 1
Deed, Petit Manan Land and Industrial Company to James Bennett, Vol. 212, Page 70, Sheriff sale
Deed, Petit Manan Land and Industrial Company to Jane M. Feigl, Vol. 213, Page 458, Lots 144–146, Section 2
Deed, Petit Manan Land and Industrial Company to John H. Flannagan, Vol. 212, Page 78, Sheriff sale
Deed, Petit Manan Land and Industrial Company to James L. Gougins, Vol. 222, Page 61, Lot 141, Section 1
Deed, Petit Manan Land and Industrial Company to L. B. Haskill, Vol. 213, Page 431, Lot 48, Section 1
Deed, Petit Manan Land and Industrial Company to International Trust Company, Vol. 213, Page 396
Deed, Petit Manan Land and Industrial Company to International Trust Company, Vol. 213, Page 402
Deed, Petit Manan Land and Industrial Company to Henry L. Kilgore et all, Vol. 224, Page 448, All Property

Deed, Petit Manan Land and Industrial Company to Edward B. Phelps, Vol. 224, Page 32, Lot 65, Section 3

Deed, Petit Manan Land and Industrial Company to Charles Woodman/McLean, Vol. 222, Page 186, Lot 6, Section 1

Deed, Petit Manan Land and Industrial Company to Charles Woodman/McLean, Vol. 222, Page 187, Lot 8, Section 1

Deed, Petit Manan Land and Industrial Company to Charles Woodman/McLean, Vol. 222, Page 188, Lot 71, Section 1

Deed, Petit Manan Land and Industrial Company to Charles Woodman/McLean, Vol. 222, Page 189, Lot 7, Section 1

Deed, Petit Manan Land and Industrial Company to Charles Woodman/McLean, Vol. 222, Page 190, Lot 5, Section 1

Deed, Petit Manan Land and Industrial Company to Charles Woodman/McLean, Vol. 222, Page 191, Lot 4, Section 1

Deed, Petit Manan Land and Industrial Company to Everard H. Greely Trustee, Vol. 221, Page 300, Lot 90, Section 1

Deed, Petit Manan Land and Industrial Company to Everard H. Greely Trustee, Vol. 221, Page 301, Lot 91, Section 1

Deed, Petit Manan Land and Industrial Company to Everard H. Greely Trustee, Vol. 221, Page 302, Lot 86, Section 1

Deed, Petit Manan Land and Industrial Company to Everard H. Greely Trustee, Vol. 221, Page 303, Lot 88, Section 1

Deed, Petit Manan Land and Industrial Company to Everard H. Greely Trustee, Vol. 221, Page 304, Lot 105, Section 1

Deed, Petit Manan Land and Industrial Company to Everard H. Greely Trustee, Vol. 221, Page 305, Hotel site, Section 1

Deed, Petit Manan Land and Industrial Company to Everard H. Greely Trustee, Vol. 221, Page 306, Lot 2, Section 1

Deed, Petit Manan Land and Industrial Company to Everard H. Greely Trustee, Vol. 221, Page 307, Lot 20, Section 1

Deed, Petit Manan Land and Industrial Company to Everard H. Greely Trustee, Vol. 221, Page 308, Lot 31, Section 1

Deed, Petit Manan Land and Industrial Company to Everard H. Greely Trustee, Vol. 221, Page 309, Lot 30, Section 1

Deed, Petit Manan Land and Industrial Company to Everard H. Greely Trustee, Vol. 221, Page 310, Lot 23, Section 1

Deed, Petit Manan Land and Industrial Company to Everard H. Greely Trustee, Vol. 221, Page 311, Lot 13, Section 1

Deed, Petit Manan Land and Industrial Company to Ella A. Fletcher, Vol. 248, Page 228, Lot 100, Section 1

Deed, JP Parker to Petit Manan Land and Industrial Company, Vol. 219, Page 150

Deed, Maine Coast Club, Robert F. Dunton Trustee to Ella H. Amory, Vol. 250, Page 3, Lot 88, Section 1

Deed, Maine Coast Club, Robert F. Dunton Trustee to George F. Barker, Vol. 250, Page 1, Lot 11, Section 1

Deed, Maine Coast Club, Robert F. Dunton Trustee to Harriet P. Corcoran, Vol. 248, Page 8, Lot 84, Section 1

Deed, Maine Coast Club, Robert F. Dunton Trustee to Edward J. Cross, Vol. 257, Page 58

Deed, Maine Coast Club, Robert F. Dunton Trustee to Eleanor H. Denniston, Vol. 248, Page 3, Lot 92, Section 1

Deed, Maine Coast Club, Robert F. Dunton Trustee to Eastern Fish & Lobster Company, Vol. 248, Page 121

Deed, Maine Coast Club, Robert F. Dunton Trustee to Charles B. Fritz, Vol. 248, Page 5, Lot 12, Section 1

Deed, Maine Coast Club, Robert F. Dunton Trustee to Henry O. Hastings, Vol. 248, Page 6, Lot 82, Section 1

Deed, Maine Coast Club, Robert F. Dunton Trustee to John M. Kilgore, Vol. 248, Page 9, Lot 30, Section 1

Deed, Maine Coast Club, Robert F. Dunton Trustee to William S. Mitchell, Vol. 257, Page 53

Deed, Maine Coast Club, Robert F. Dunton Trustee to L. M. Palmer, Vol. 250, Page 6, Lot 90, Section 1

Deed, Maine Coast Club, Robert F. Dunton Trustee to Joseph W. Paul, Vol. 248, Page 2, Lot 81, Section 1

Deed, Maine Coast Club, Robert F. Dunton Trustee to W. J. Sperl, Vol. 250, Page 4, Lot 106, Section 1

Deed, Maine Coast Club, Robert F. Dunton Trustee to W. J. Sperl, Vol. 250, Page 5, Lot 93, Section 1

Deed, Maine Coast Club, Robert F. Dunton Trustee to J. Frank Turner, Vol. 248, Page 1, Lot 89, Section 1

Deed, Maine Coast Club, Robert F. Dunton Trustee to Marianne Whitney, Vol. 250, Page 2, Lot 86, Section 1

Deed, Maine Coast Club, Robert F. Dunton Trustee to Duane B. Williams, Vol. 248, Page 10, Lot 31, Section 1

Deed, Maine Coast Club, Robert F. Dunton Trustee to Harry Goddard, Vol. 250, Page 7, Lots 59–64, Section 1

Deed, Maine Coast Club, Robert F. Dunton Trustee to G. Horton Glover, Vol. 250, Page 8, Lots 121–124, Section 3

Deed, Maine Coast Club, Robert F. Dunton Trustee to George L. Hooper, Vol. 250, Page 9, Hotel Site on Yellow Birch Head

Deed, Maine Coast Club, Robert F. Dunton Trustee to Mary D.B. Hooper, Vol. 250, Page 10, Lot 134–136, Section 3

Deed, Maine Coast Club, Robert F. Dunton Trustee to Lester Bailey, Vol. 250, Page 11, Lot 2, Section 2

Deed, Maine Coast Club, Robert F. Dunton Trustee to Lester Bailey, Vol. 250, Page 12, Lots 185, 240–241, 10' of 186, Section 2

Deed, Maine Coast Club, Robert F. Dunton Trustee to Frances Shaw, Vol. 250, Page 13, Hotel Site, Section 1

Deed, Maine Coast Club, Robert F. Dunton Trustee to William Bowring Windeler, Vol. 250, Page 14, Lot 98, Section 1

Deed, Maine Coast Club, Robert F. Dunton Trustee to Herbert Jaques, Vol. 250, Page 15, Lot 99, Section 1

Deed, Maine Coast Club, Robert F. Dunton Trustee to Clarence H. Crocker, Vol. 250, Page 16, Lot 367, Section 3

Deed, Maine Coast Club, Robert F. Dunton Trustee to Herbert Carlton. Vol. 250, Page 17, Lot 57–65

Deed, Maine Coast Club, Robert F. Dunton Trustee to LM Palmer, Vol. 317, Page 157, Lot 90, Section 1

Deed, Maine Coast Club, Robert F. Dunton Trustee to Mary B. Palmer (in trust), Vol. 250, Page 16

Deed, Maine Coast Club, Robert F. Dunton Trustee to Janet C. Sheppard, Vol. 317, Page 158, Hotel Site, Section 1

Deed, Maine Coast Club, Robert F. Dunton Trustee to Lewis M. Palmer, Vol. 250, Page 6

Deed Mortgage, John M. Kilgore to Alonzo Smith, Vol. 281, Page 3, Lots 38 & 134, Section 1, July 8, 1907

Deed, Alonzo Smith to John Kilgore, Vol. 279, Page 53, Lots 38 & 134, Section 1, July 8, 1907

Deed, John Kilgore Mortgage Transfer, from Alonzo Smith to Byron Smith, Vol. 299, Page 392, December 11, 1916

Deed, John Kilgore Mortgage Transfer, from Byron Smith to Liberty National Bank, Vol. 299, Page 392, January 22, 1920

Deed, Mortgage Discharge, John Kilgore by Liberty National Bank, Vol. 362, Page 315, January 2, 1926

Deed, Maine Coast Club, John M. Kilgore Trustee to John Kilgore, Vol. 294, Page 496, Lot 105

Deed, Maine Coast Club, John M. Kilgore Trustee to G. Horton Glover, Vol. 377, Page 136, Lot 125, Section 3, July 9, 1928

Deed, Maine Coast Club, John M. Kilgore Trustee to Tabitha Fitz-Gerald, Vol. 389, Page 125, Lot 91, Section 1, April 9, 1930

Deed, Alonzo Smith to John M. Kilgore, Vol. 279, Page 53, Lots 38 & 134, Section 1

Deed, Clarence H. Crocker (trustee for Mary B. Palmer) to John M. Kilgore, Vol., Page 127, Lots 367–368, 381–383, Section 3, June 10, 1926

Deed, Robert A. Davis to John M. Kilgore, Vol. 369, Page 71, Lot 31, Section 3, March 31, 1926

Deed, Harry W. Goddard to John M. Kilgore, Vol. 369, Page 129, Lots 59–64, Section 1, June 7, 1926

Deed, Nellie B. Palmer to John M. Kilgore, Vol. 369, Page 95, Lot 90, Section 1, May 5, 1926

Deed, Inhabitants of Steuben to John M. Kilgore, Vol. 369, Page 19, Lots 98–99, Section 1, Lots 37, 71, 75, 91, 100–103, 113, 142, 144–146, 211, Section 2, Lots 125, 147, 173, 367, 569, Section 3, Lots 117, 137, 157–158, 185, 199, 245, Section 4, January 6, 1926

Deed, Duane B. Williams to John M. Kilgore, Vol. 369, Page 126, Lot 31, Section 1, June 9, 1926

Deed, Hastings & Son Publishing Company to John M. Kilgore, Vol. 369, Page 41, Lot 42, 85, Section 1, January 29, 1926

Deed, George T. Thatcher to John M. Kilgore, Vol. 369, Page 142, Lot 10, Section 1, June 24, 1926

Deed, Allan W. McLean to Minnie Casey, Vol. 387, Page 576, Lots 4–8, 71, Section 1, September 29, 1932

Deed, Petit Manan in Maine Inc. to Emma M. Sibley, Vol. 389, Page 103, Lot 76 & 77, Section 1, March 1, 1930

Deed, Petit Manan in Maine Inc. to Leon F. Lawrence, Vol. 389, Page 531, Lots 145–150, Section 3, January 25, 1930

Deed, Petit Manan in Maine Inc. to Robert Simpson, Vol. 389, Lots 113–120, Section 3, December 1930

Deed, Petit Manan in Maine Inc. to Eleanor G. Daniels, Vol. 437, Page 252, Lots 13, 17, 20, 23, 30–31, 38, 53, 60, 62–64, 71, 81, 90, 99–101, 134, 141, Hotel site, 3–6, Section 1, Lots 37, 71, 75, 91, 100–103, 113, 142, 144–146, 186, 211, Section 2, Lots 31, 147, 173, 367–368, 381–383, 569, Section 3, Lots 117, 137, 157–158, 185, 199, 245, 247, 289, Section 4, January 24, 1941

Deed, Inhabitants of Steuben to Petit Manan in Maine Inc., Vol. 428, Page 468, Lots 13, 17, 20, 23, 30–31, 38, 53, 60, 62–64, 71, 81, 90, 99–101, 134, 141, Hotel site, 3–6, Section 1, Lots 37, 71, 75, 91, 100–103, 113, 142, 144–146, 186, 211, Section 2, Lots 31, 147, 173, 367–368, 381–383, 569, Section 3, Lots 117, 137, 157–158, 185, 199, 245, 247, 289, Section 4

Deed, Eleanor G. Daniels to Mary Ellen Chase, Vol. 437, Page 599, Lots 121–125, Section 3, December 18, 1941

Deed, Eleanor G. Daniels to William R. Mague, Vol. 494, Page 335, March 6, 1950

Deed, Eleanor G. Daniels to Milton A. Hutchins and William F. Sawyer, Vol. 476, Page 441, Lots 3–4, Section 1, July 16, 1948

Deed, Eleanor G. Daniels to Agnes E. Dardis, Vol. 486, Page 314, May 16, 1949

Deed, Eleanor G. Daniels to Lewis S. Loud, Vol. 486, Page 313, May 16, 1949

Deed, Eleanor G. Daniels to Helen D. Perkins, Vol. 449, Page 473, June 28, 1949

Deed, Eleanor G. Daniels to Milton T. Stauffer and Marjorie H. Stauffer, Vol. 449, Page 43, May 5, 1943

Deed, Eleanor G. Daniels to Petit Manan in Maine, Inc., Vol. 494, Page 334, March 7, 1950

Deed, Eleanor G. Daniels to Katahdin Council Inc., BSA, Vol. 494, Page 90, 360 acres including the clubhouse and two log cabins, January 31, 1950

Deed, Eleanor G. Daniels to Henry D. Booth, Vol. 494, Page 127, January 31, 1950

Deed, Eleanor G. Daniels to Maine Department of Inland Fisheries and Game, Vol. 494, Page 131, 500 acres near Bear Cove, February 28, 1950

Deed, Eleanor G. Daniels to Anna Denniston, Vol.494, Page 105, lot next to Bairnbrae, June 28, 1949

Deed, Eleanor G. Daniels to Anna Denniston, Vol. 494, Page 104, January 31, 1950

Deed, Eleanor G. Daniels to George P. Jacobs, Vol. 526, Page 58, 3 acres near Hilton lot, September 12, 1949

Deed, Katahdin Council Inc., BSA to Eleanor G. Daniels, Vol. 555, Page 527, Return of land, May 27, 1958

Deed, Maine Department of Inland Fisheries and Game to Eleanor G. Daniels, Vol. 555, Page 330, Return of land, March 26, 1958

Deed, Eleanor G. Daniels to Petit Manan Industries, Vol. 556, Page 222, Land from BSA and Maine Inland Fisheries, February 17, 1959

Deed, Town of Steuben to William R. Mague, Vol. 840, Page 98, Lot 85, Section 2, May 17, 1974

Deed, Petit Manan Industries, Inc. to William R. Mague, Vol. 895, Page 182, Mague Homestead, August 15, 1975

Deed, Petit Manan Industries, Inc. to William R. Mague, Vol. 895, Page 186, All the Shores and Flats, August 15, 1975

Corrective Deed, Petit Manan Industries, Inc. to William R. Mague, Vol. 901, Page 243, To correct Aug 15, 1975, October 14, 1975

Deed, William R. Mague to Mague Inc., Vol. 1168, Page 8, Mague Homestead, Dec. 31, 1981

Deed, Mague Inc., to Peter Dardis, Vol. 1431, Page 206, November 12, 1986

Deed, Estate of Glenore O. Coblens to Frances C. Mague, Vol. 1452, Page 29, Beach lot in Section 1, June 18, 1987

Deed, Estate of Glenore O. Coblens to Frances C. Mague, Vol. 1613, Page 268, Tax lot 003–004, Section 3, July 24, 1989

Deed, Frances C. Mague to Frances C. Mague Living Trust, Vol. 3648, Page 267, Tax lot 003–003, Section 3, June 30, 2010

Deed, Frances C. Mague to Frances C. Mague Living Trust, Vol. 3648, Page 269, Tax lot 003–004, Section 3, June 30, 2010

Deed, Frances C. Mague to Frances C. Mague Living Trust, Vol. 3648, Page 271, Beach lot in Section 1, June 30, 2010

Deed, Milton T. Stauffer and Marjorie H. Stauffer to Richard C. Coblens, Vol. 567, Page 465, June 16, 1960

Deed, Richard C. Coblens Estate to Glenore O. Coblens, Vol. 1304, Page 102, November 21, 1984

Deed, Robert Simpson heirs, Jessie S. Laurson, Lillian S. Allen, Robert K. Welliver to Eleanor G. Daniels, Vol. 494, Page 552, Lots 113–114, 116–120, Section 3, November 1, 1950

Deed, W. T. Howard to Eleanor G. Daniels, Vol. 499, Page 95, Lot 115, Section 3, November 15, 1950

Deed, Petit Manan Industries Inc. to Eleanor G. Daniels, Vol. 664, Page 349–350, Lots 113–120, Section 3, September 29, 1969

Deed, Town of Steuben to David H. Leighton & Spencer Leighton, Vol. 486, Page 2, Tax lot 003–003, Section 3, January 20, 1948

Deed, Spencer H. Leighton & Corinne Dionne to Eleanor G. Daniels, Vol. 558, Page 221, Tax lot 003–003, Section 3, February 4, 1959

Deed, Eleanor G. Daniels to Robert M. Mitchell, Vol. 700, Page 123, November 27, 1970

Deed, Robert Mitchell to Robert M. Mitchell and Mary Ann Mitchell, Vol. 711, Page 31, March 4, 1971

Deed, Robert M. Mitchell and Mary Ann McGeough to Frances C. Mague, Vol. 1357, Page 197, Tax lot 003–003, Section 3, October 24, 1985

Deed, Frances C. Mague to Frances C. Mague and Glenore O. Coblens, Vol. 1357, Page 198, Tax lot 003–003, Section 3, November 5, 1985

Deed, John M. Kilgore to John M. Kilgore, Eleanor Denniston, Lester V. Bailey, Vol. 298, Page 456, Lot 130, Section 1, January 6, 1914

Deed, Henry D. Booth Jr. to All Souls by the Sea, Vol. 567, Lot 481, All land north of lots 135–136 in Section 1, August 28, 1959

Deed, Petit Manan Industries to All Souls by the Sea, Vol. 626, Page 505, Additional land near graveyard, August 26, 1966

Deed, Hame Corp. to All Souls by the Sea, Vol. 1357, Page 105, Lot 131, Section 1, October 8, 1985

Deed, All Souls by the Sea to Margaret D Kershaw, Vol. 580, Page 448, Lot near Bathing Pool, Section 1, August 21, 1962

Deed, All Souls by the Sea to Edward E. Denniston, Vol. 626, Page 469, All land north of lots 135–136 in section 1, August 9, 1966

Deed, Eleanor H. Denniston to Anna H. Denniston, Vol. 387, Page 496, Lot 92, Section 1, May 19, 1932

Deed, Anna H. Denniston to Trustees Edward E. Denniston, Margaret D. Kershaw, Leonora D. Booth, Kathleen D. Kent, Vol. 778, Page 27, Lot 92, Section 1, May 19, 1932

Deed, David H. Leighton and Spencer Leighton to Dorothy S. Vaughan, Vol. 494, Page 27, Lot 57–58, 66, Section 1, October 31, 1949

Deed, David H. Leighton and Spencer Leighton to Dorothy S. Vaughan, Vol. 486, Page 479, Lot 65, Section 1, September 3, 1949

Deed, Dorothy S. Vaughan to Edward E. Denniston, Jr., Vol. 624, Page 212, Lots 57–58, 65–66, Section 1, November 15, 1966

Deed, Helen D. Perkins and Julia W. Williamson to Anna H. Denniston, Vol. 593, Page 76, lot below 92, Section 1, February 28, 1950

Deed, Helen D. Perkins to Anna H. Denniston, Vol. 593, Page 28, lot near 92, Section 1, February 28, 1950

Deed, Eleanor G. Daniels to Anna H. Denniston, Vol. 494, Page 104, lot below 92 to section 2, Lots 93–106, Section 1, January 31, 1950

Deed, Eleanor G. Daniels to Anna H. Denniston, Vol. 494, Lot 105, Section 1, June 28, 1949

Deed, Helen D. Perkins to Anna H. Denniston, Vol. 593, Page 28

Deed, Edward E. Denniston, Jr., Trustee to Bairnbrae Corp., Vol. 1729, Page 314, Lot 92, Section 1, October 16, 1991

Deed, Edward E. Denniston, Jr., Mary B. Brush, Donald W. Kent, Jr., William E. Kershaw to Bairnbrae Corp., Vol. 1829 Page 5, Lots 94–105, Section 1, November 24, 1992

Deed, Edward E. Denniston, Jr., to Bairnbrae Corp., Vol. 1846, Page 131, Overs Point (lots 1–2), Section 1, April 30, 1993

Deed, Bairnbrae Corp. to Over's Cove Corp., Vol. 2955, Page 122, Overs Point (lots 1–2), Section 1, April 22, 2004

Deed, Edward E. Denniston, Jr., to Barbara D. Snapp, Evans E. Denniston, Michael M. Denniston, Vol. 1846, Page 133 all land north of lot 135 excepting lands already reserved, April 30, 1993

Deed, Michael M. Denniston to Barbara D. Snapp and Evans E. Denniston, Vol. 1881, Page 141, all land north of lot 135 excepting lands already reserved, August 18, 1993

Deed, Barbara D. Snapp, Evans E. Denniston, Elizabeth R. Denniston to Great Auk Land Trust, Vol. 3042, Page 301, Lots 6–9, 16–23, 72–79, Sand Point, Section 1, August 5, 2005

Deed, Barbara D. Snapp and Evans E. Denniston to Craig P. Snapp and Barbara D. Snapp Trustees for Snapp Family Trust, Vol. 3042, Page 303, Lot 15, area between 15 and 141, 32–33, 50–51, 140–141, Section 1, August 5, 2005

Deed, Barbara D. Snapp and Evans E. Denniston to Michael M. Denniston, Vol. 3042, Page 305, Lots 10–12, northern half of land between 12 and 13, Section 1, August 5, 2005

Deed, Barbara D. Snapp and Evans E. Denniston to Evans E. Denniston, Vol. 3042, Page 307, Southern half of land between 12 and 13, Lots 13–14, 30–31, Section 1, August 5, 2005

Deed, Bairnbrae Corp. to James D. Kovacs and Margaret D. Kovacs, Vol. 3677, Page 131, September 15, 2010

Deed, James D. Kovacs and Margaret D. Kovacs to Bairnbrae Corp., Vol. 3678, Page 63, Lots 3–4, Section 1, September 20, 2010

Deed, Bairnbrae Corp., to Downeast Coastal Conservancy, Vol. 3725, Page 39, lots 3–4, Section 1, March 8, 2011

Deed, Bairnbrae Corp., to Downeast Coastal Conservancy, Vol. 3849, Page 305, lot 5, corrective deed to add lot 5, Section 1, May 15, 2012

Deed, Snapp Family Trust to Friends of Maine Coastal Islands National Wildlife Refuge, Vol. 4280, Page 86, Lot 15, area between 15 and 141, Lots 140–141, Section 1, August 1, 2016

Deed, Tabitha Fitz-Gerald to Helen D. Perkins, Vol. 471, Page 455, Lot 91, Section 1, August 31, 1946

Deed, Petit Manan in Maine Inc. to Helen D. Perkins, Vol. 486, Page 430, Lot 90, Section 1, September 7, 1949

Deed, Helen D. Perkins to Thomas M. & Amy B. McIlvaine, Vol. 598, Page 351, Lots 90–91, Section 1, December 7, 1964

Deed, Thomas M. & Amy B. McIlvaine to Thomas M., Elizabeth T., Lee D. McIlvaine, and Mary M. Selin, Vol. 1871, Page 29

Deed, Thomas M., Elizabeth T., Lee D. McIlvaine, and Mary M. Selin to Lee D. McIlvaine trustee McIlvaine Family Trust, Vol. 3706, Page 11, Lots 90–91, Section 1, December 29, 2010

Deed, Robert G. Feeney & Frances M. Feeney to Charles A. Sullivan & Dawna P. Sullivan, Vol. 2104, Page 195, Lots 89, parts of 87 and 88, Section 1, August 5, 1996

Deed, Robert G. Feeney & Frances M. Feeney to Charles A. Sullivan & Dawna P. Sullivan, Vol. 2279, Page 107, Lot Line adjustment, September 21, 1998

Deed, Charles A. Sullivan & Dawna P. Sullivan to Dawna P. Sullivan Trust, Vol. 2500, Page 1, Lots 89, parts of 87 and 88, Section 1, March 19, 2001

Deed, Badger & Manley to Lillian E. Cassens, Vol. 240, Page 224, Lot 87, Section 1, March 7, 1901

Deed, Lillian E. Cassens to Ella Howard Amory, Vol. 247, Page 425, Lot 88, Section 1, November 3, 1902

Deed, Estate of Ella Howard Amory to Frank T. Whitney, Vol. 367, Page 493, Lot 86, Section 1, October 16, 1925

Deed, Frank T. Whitney to George Dobbin Penniman, Vol. 426, Page 434, May 23, 1939

Deed, George Dobbin Penniman to Frank T. & Alexandra Dushane Whitney, Vol. 425, Page 432, May 23, 1939

Deed, Alexandra Dushane Whitney to Margaret Speaks Pearl, Vol. 479, Page 108, Lots 86–88, Section 1, May 26, 1947

Deed, Carmen G. Whitten, Maurice H. Whitten, John F. Whitten to Margaret Speaks Pearl, Vol. 501, Page 213, Lots 85, 89, Section 1, May 25, 1950

Deed, Margaret Speaks Pearl to Edwin H. Craig and Margaret Smythe, Vol. 528, Page 169, Lots 85–89, Section 1, September 1, 1955

Deed, Edwin H. Craig and Elizabeth S. Craig to Margaret Smythe, Vol. 605, Page 218, Lots 85–89, Section 1, July 31, 1965

Deed, Margaret Smythe to Robert G. Feeney and Georgiana Feeney, Vol. 708, Page 294, Lots 85–89, Section 1, March 20, 1971

Deed, Robert G. Feeney to Robert G. Feeney and Frances M. Feeney Trustees Feeney Family Trust, Vol. 1687, Page 155, Lots 85–89, Section 1, March 5, 1991

Deed, Margaret S. Paul to Henry D. Booth, Jr., and Leanora D. Booth, Vol. 385, Page 20, Lot 81, Section 1, August 30, 1929

Deed, Pennsylvania Company for Banking and Trusts under will of Henry O. Hastings to Henry D. Booth, Jr., and Leanora D. Booth, Vol. 496, Page 126, Lot 82, Section 1, July 21, 1950

Deed, Town of Steuben to Henry D. Booth, Jr., and Leanora D. Booth, Vol. 494, Page 420, Lot 83, Section 1, August 18, 1950

Deed, Eleanor G. Daniels to Henry D. Booth, Jr., and Leanora D. Booth, Vol. 494, Page 127, Hotel site, Lot 62, Section 1, January 31, 1950

Deed, Carmen G. Whitten, Maurice H. Whitten, John F. Whitten to Henry D. Booth, Jr., and Leanora D. Booth, Vol. 501, Page 210, Lot 42, Section 1, May 25, 1950

Deed, Town of Steuben to Henry D. Booth, Jr., Vol. 569, Page 132, Lot 61, Section 1, July 15, 1947

Deed, Louise H. Whitten to Henry D. Booth, Jr., and Leanora D. Booth, Vol. 501, Page 212, Lot 84, Section 1, September 27, 1950

Deed, Town of Steuben to Spencer H. Leighton, Vol. 486, Page 2, Lots 11, 55, Section 1, January 20, 1948

Deed, Spencer H. Leighton to Henry D. Booth, Jr., Vol. 558, Page 196, Lots 11, 55, Section 1, January 30, 1959

Deed, Petit Manan Industries to Henry D. Booth, Jr., Vol. 569, Page 326, North part of Point, Section 1, February 17, 1959

Deed, Henry D. Booth, Jr., and Leanora D. Booth to Leanora D. Booth, Vol. 563, Page 418, Lots 81–83, hotel site, 38–45, 61–62, 84, 133, Section 1, February 13, 1960

Deed, Edward E. Denniston, Jr., to Leanora D. Booth, Vol. 664, Page 384, Lots 57–58, 65–66, Section 1, October 23, 1969

Deed, Henry D. Booth, Jr., and Leanora D. Booth to Hame Corp., Vol. 904, Page 151, Lot 81–83, Hotel Site, 11, 38–45, 55, 61–62, 84, 133, Section 1, October 31, 1975

Deed, Edward E. Denniston, Jr., to Henry D. Booth, Vol. 1291, Page 261, Lots 34, 46–49, 80, 136–139, Section 1, July 30, 1984

Deed, Edward E. Denniston, Jr., to Margot V. Booth, Vol. 1844, Page 4, Lots 34–37, 46–49, 80, 136–139, Corrective deed for above, Section 1, April 30, 1993

Deed, Margot V. Booth to Ann D. Booth, Claire B. Parker, Katherine B. Baum, Andrew P. Booth, James H. Booth, Vol. 3706, Page 309, Lots 34–37, 46–49, 80, 136–139, Section 1, December 27, 2010

Deed, Ann D. Booth, Claire B. Parker, Katherine B. Baum, Andrew P. Booth, James H. Booth to Hame Corp., Vol. 3846, Page 88, Lot 80, Section 1, April 23, 2012

Deed, Ann D. Booth, Claire B. Parker, Katherine B. Baum, James H. Booth to Andrew P. Booth, Vol. 3919, Page 276, Lots 34–37, 46–49, 136–139, Section 1, December 18, 2012

Deed, Margaret K. (Brown) Letterman to Wendy K. Brown, Vol. 1210, Page 294, August 16, 1982

Deed, Oliver L. Page to Gardiner H. Dutton, Vol. 242, Page 306, Lot 112, Section 1, October 15, 1900

Deed, John M. Kilgore to Gardiner H. Dutton, Vol. 366, Page 28, extra land next to 112, Section 1, January 23, 1925

Deed, Gardiner H. Dutton to Effie M. Dutton, Vol. 413, Page 453, Lot 112+, Section 1, June 2, 1936

Deed, Marion B. Dutton Estate, Vol. 868, Page 215, Lot 112+, Section 1, one-third interest in cottage lot at Petit Manan Point

Deed, Town of Steuben to T. Dudley Drew-Bear, Vol. 358, Page 204, Lots 93, 106, Section 1, July 15, 1924

Deed, Kenneth K. Sperl to T. Dudley Drew-Bear, Vol. 372, Page 476, Lots 93, 106, Section 1, September 15, 1925

Deed, Martha L. Sperl to T. Dudley Drew-Bear, Vol. 373, Page 173, Lots 93, 106, Section 1, April 4, 1925

Deed, T. Dudley Drew-Bear to Jessie Drew-Bear, Vol. 379, Page 573, Lots 93, 106, Section 1, May 31, 1929

Deed, Jessie Drew-Bear to William Brewster Taber, Vol. 399, Page 242, Lot 93, 106, Section 1, October 24, 1933

Deed, William Brewster Taber to Elizabeth Roberta Puffer, Lots 93, 106, Section 1, November 20, 1936

Deed, Roberta Puffer to John Morrison and Irene Puffer Morrison, Vol. 516, Page 4, Lots 93, 106, Section 1, June 26, 1951

Deed, Homer E. Morrison to Irene P. Morrison, Vol. 732, Page 95, Lots 93, 106, Section 1, February 27, 1970

Deed, Irene P. Morrison to Homer P. Morrison, Priscilla E. Diehl, John A. Morrison, Jr., Bonnie Mae King, Vol. 1354, Page 191, Lots 93, 106, Section 1, December 9, 1976

Deed, William R. Mague to Agnes E. Dardis, Vol. 494, Page 391, March 8, 1950

Deed, Agnes E. Dardis to Thomas D. & Mary C. Dardis, Vol. 528, Page 428, February 9, 1956

Deed, William R. Mague to Thomas D. & Mary C. Dardis, Vol. 627, Page 217, September 12, 1968

Deed, Thomas D. Dardis & Mary C. Dardis to Dardis Corp., Vol. 1305, Page 133, October 19, 1984

Deed, Dardis Corp. to Dardis Family Real Estate Trust—Trustees—Peter Dardis, Kathleen Dardis Noe, Margaret D. Kovacs, Regina Marie Dardis, Vol. 3043, Page 63, June 28, 2005

Deed, Mague Inc.to Peter Dardis, Vol. 1431, Page 206, November 12, 1986

Deed, Peter Dardis to Peter & Roseann Dardis, Vol. 1528, Page 346, August 12, 1988

Deed, William R. Mague to Stuart Phelps, Vol. 494, Page 407, March 9, 1950

Deed, Stuart E. Phelps to Bruce R. Phelps, Vol. 624, Page 425, May 5, 1967

Deed, Bruce R. Phelps to Walter R. & Suzanne Armstrong, Vol. 1125, Page 37, December 15, 1980

Deed, Walter R. Armstrong to Suzanne Armstrong, Vol. 2041, Page 194, November 15, 1995

Deed, William R. Mague to Harriet E. Heath, Vol. 514, Page 270, 1st lot, September 14, 1953

Deed, William R. Mague to Harriet E. Heath, Vol. 542, Page 22, 2nd lot, August 2, 1966

Deed, William R. Mague to Harriet E. Heath, Vol. 645, Page 496, 3rd lot, September 12, 1968

Deed, Harriet E. Heath to Douglas H. Heath and Harriet E. Heath, Vol. 1541, Page 324, October 18, 1988

Deed, William Mague to Stuart E. Phelps, Vol. 494, Page 407, March 9, 1950

Deed, Stuart E. Phelps to Stuart E. Phelps, Jr., Vol. 624, Page 424, May 22, 1967

Deed, Stuart E. Phelps, Jr., to Cheryl M and Richard L. Moisan, Vol. 1139, Page 134, May 28, 1981

Deed, William R. Mague to Verdi L & Margaret H. Reusser, Vol. 569, Page 399, July 28, 1954

Deed, Verdi L. Reusser and Margaret H. Reusser to Verdi L. Reusser, Margaret H. Reusser, David H. Reusser, Jonathan W. Reusser, Madelyn E. Reusser, Vol. 1868, Page 122, January 27, 1981

Deed, Margaret H. Reusser, David H. Reusser, Jonathan W. Reusser, Madelyn E. Reusser to Jonathan W. Reusser, Madelyn E. Reusser, Christina M. Reusser, Lucas J. Reusser, Megan C. Reusser, Vol. 2272, Page 290, April 21, 1998

Deed, G. Horton Glover to Milton T. Stauffer, Vol. 410, Page 310, Lots 121–125, Section 3, July 22, 1936

Deed, Milton T. Shauffer to Eleanor G. Daniels, Vol. 419, Page 353, Lots 121–125, Section 3, August 11, 1937

Deed, Eleanor G. Daniels to Mary Ellen Chase, Vol. 437, Page 599, Lots 121–125, Section 3, December 18, 1941

Deed, Mary Ellen Chase to Leslie S. Pearl, Vol. 537, Page 216, Lot 121–125, Section 3, August 25, 1955

Deed, Leslie S. Pearl to Jane A. Skinner, Vol. 608, Page 520, Lots 121–125, Section 3, September 29, 1965

Deed, Jane A. Skinner to Peter C. Huycke and Jane S. Huycke, Vol. 1556, Page 259, Lots 121–125, Section 3, January 13, 1989

Deed, Peter C. Huycke to Peter C. Huycke and Lynda Gentile-Huycke, Vol. 3345, Page 81, Lots 121–125, Section 3, October 19, 2007

Deed, Peter C. Huyche and Lynda Gentile-Huycke to Mark V. Tremallo, Vol. 3600, Page 266, Lots 121–125, Section 3, January 5, 2010

Deed, Mark V. Tremallo to Wellner A. Poppe, Vol. 3890, Page 79, one-half interest in property, October 4, 2012

Deed, Mark V. Tremallo to Mark V. Tremallo Residence Trust, Vol. 3912, Page 60, December 13, 2012

Deed, Wellner A. Poppe to Wellner A. Poppe Residence Trust, Vol. 3912, Page 58, December 13, 2012

Deed, Petit Manan Industries to Paul Wehr, Vol. 743. Page 182, January 12, 1972

Deed, Paul Wehr to The Friends of Maine's Seabird Islands, Vol. 4216 Page 203, December 8, 2015

Deed, Petit Manan Industries to Donald B. and Janet M. Duncan, Vol. 674, Page 503, October 13, 1969

Deed, Julia Duncan Rinehart to Michael and Gertrude M. Sloan, Vol. 2109, Page 138, August 26, 1996

Deed, Michael and Gertrude M. Sloan to Gertrude M. Sloan, Vol. 2251, Page 80, May 6, 1998

Deed, Gertrude M. Sloan to Barbara W. Parson Trust, Vol. 3419, Page 12, June 4, 2008

Deed, Barbara W. Parson Trust to Wendi A. Parson and Kent R. Parson, Vol. 4173, Page 92, July 21, 2015

Deed, Wendi A. Parson to Kent R. Parson, Vol. 4234, Page 11, February 17, 2016

Deed, Petit Manan Industries to John J. and Marjorie M. Scully, Vol. 635, Page 213, March 22, 1967

Deed, John J. and Marjorie M. Scully to Scully Revocable Trust, Vol. 1935, Page 207, March 11, 1994

Deed, Scully Revocable Trust to Georges Real Estate Trust, Vol. 2591, Page 28, December 31, 2001

Deed, Julia Duncan Rinehart to George Scully, Vol. 2374, Page 259, July 19, 1999

Deed, George M. and Gesele Scully to The United States of America, Vol. 4420, Page 274, November 27, 2017

Deed, Kent Parsons to The United States of America, Vol. 4427, Page 166, December 15, 2017

Deed, Friends of Maine Coastal Islands National Wildlife Refuge to The United States of America, Vol. 4394, Page 308, September 1, 2017.